ANALYSES FOR DURABILITY AND SYSTEM DESIGN LIFETIME

An important issue in engineering design is a system's design lifetime. Economists study durability choice problems for consumer goods, but seldom address lifetime problem(s) of complex engineering systems. The issues for engineering systems are complex and multidisciplinary and require an understanding of the "technicalities of durability" and the economic implications of the marginal cost of durability and of value maximization. Commonly the design lifetime for infrastructure is set between 30 and 70 years, with limited rationale. Satellite lifetimes are also assigned with limited analysis. This book provides a systemic qualitative and quantitative approach to these problems, addressing first the technicality of durability, second the marginal cost of durability, and third the durability choice problem for complex engineering systems with network externalities (competition and market uncertainty) and obsolescence effects (technology evolution). Because the analyses are system-specific, a satellite example is used to illustrate the essence and provide a quantitative application of these analyses.

Dr. Joseph H. Saleh is an Assistant Professor of Aerospace Engineering at the Georgia Institute of Technology. He received his Ph.D. from the Department of Aeronautics and Astronautics at MIT and served as the Executive Director of the Ford–MIT Alliance. His research focuses on issues of design lifetime and how to embed flexibility in the design of complex engineering systems in general and in aerospace systems in particular. Dr. Saleh is the author or co-author of 50 technical publications and the recipient of numerous awards for his teaching and research contributions. He served as a technical consultant to NASA's Jet Propulsion Laboratory and has collaborated on research projects with various aerospace companies.

CAMBRIDGE AEROSPACE SERIES

Editors: Wei Shyy and Michael J. Rycroft

1. J. M. Rolfe and K. J. Staples (eds.): *Flight Simulation*
2. P. Berlin: *The Geostationary Applications Satellite*
3. M. J. T. Smith: *Aircraft Noise*
4. N. X. Vinh: *Flight Mechanics of High-Performance Aircraft*
5. W. A. Mair and D. L. Birdsall: *Aircraft Performance*
6. M. J. Abzug and E. E. Larrabee: *Airplane Stability and Control*
7. M. J. Sidi: *Spacecraft Dynamics and Control*
8. J. D. Anderson: *A History of Aerodynamics*
9. A. M. Cruise, J. A. Bowles, C. V. Goodall, and T. J. Patrick: *Principles of Space Instrument Design*
10. G. A. Khoury and J. D. Gillett (eds.): *Airship Technology*
11. J. Fielding: *Introduction to Aircraft Design*
12. J. G. Leishman: *Principles of Helicopter Aerodynamics, 2nd Edition*
13. J. Katz and A. Plotkin: *Low Speed Aerodynamics, 2nd Edition*
14. M. J. Abzug and E. E. Larrabee: *Airplane Stability and Control: A History of the Technologies That Made Aviation Possible, 2nd Edition*
15. D. H. Hodges and G. A. Pierce: *Introduction to Structural Dynamics and Aeroelasticity*
16. W. Fehse: *Automatic Rendezvous and Docking of Spacecraft*
17. R. D. Flack: *Fundamentals of Jet Propulsion with Applications*
18. E. A. Baskharone: *Principles of Turbomachinery in Air-Breathing Engines*
19. Doyle D. Knight: *Elements of Numerical Methods for High-Speed Flows*
20. C. Wagner, T. Huettl, and P. Sagaut: *Large-Eddy Simulation for Acoustics*
21. D. Joseph, T. Funada, and J. Wang: *Potential Flows of Viscous and Viscoelastic Fluids*
22. W. Shyy, Y. Lian, H. Liu, J. Tang, and D. Viieru: *Aerodynamics of Low Reynolds Number Flyers*
23. J. Saleh: *Analyses for Durability and System Design Lifetime*

Analyses for Durability and System Design Lifetime

A MULTIDISCIPLINARY APPROACH

Joseph H. Saleh

Georgia Institute of Technology

CAMBRIDGE UNIVERSITY PRESS
Cambridge, New York, Melbourne, Madrid, Cape Town, Singapore, São Paulo, Delhi

Cambridge University Press
32 Avenue of the Americas, New York, NY 10013-2473, USA

www.cambridge.org
Information on this title: www.cambridge.org/9780521867894

First published 2008

Printed in the United States of America

A catalog record for this publication is available from the British Library.

Library of Congress Cataloging in Publication Data

Saleh, Joseph H., 1971–
Analyses for durability and system design lifetime : a multidisciplinary approach / Joseph H. Saleh.
    p.   cm. – (Cambridge aerospace series)
Includes bibliographical references and index.
ISBN 978-0-521-86789-4 (hardback)
1. Reliability (Engineering)   2. Service life (Engineering)   3. System design.   4. Reliability
(Engineering) – Economic aspects.   I. Title.   II. Series.
TA169.S235   2007
620′.00452 – dc22   2007024374

ISBN   978-0-521-86789-4 hardback

To Carl, Mia, Jihad, Abu Ali, Na'ama, and Michal

That they may find fulfillment in a peaceful, diversely rich, and prosperous Middle East

To The Reader

I know that, despite my care, nothing will be easier than to criticize this work if anyone ever thinks of criticizing it. I think those who want to regard it closely will find, in the entire work, a mother thought that so to speak links all its parts. But the diversity of the objects I had to treat is very great, and whoever undertakes to oppose an isolated fact to the sum of facts I cite or a detached idea to the sum of ideas will succeed without difficulty. I should therefore wish that one do me a favor of reading me in the same spirit that presided over my work, and that one judges this work by the general impression it leaves, just as I myself decided, not by such and such a reason, but by the mass of reasons.

Alexis de Tocqueville, *Democracy in America*, 1835

Contents

Preface   *page* xi

1   **Introduction: On Time**   1

   1.1   Sundials and human time   1
   1.2   Time and human artifacts   5
   1.3   Two broad categories of questions regarding durability   5
   1.4   Why the interest in product durability and system design
       lifetime?   8
   1.5   Book organization   10

2   **To Reduce or to Extend Durability? A Qualitative Discussion**
   of Issues at Stake   14

   2.1   Introduction   14
   2.2   Nomenclature: Durability and design lifetime – A matter of
       connotation   15
   2.3   To reduce or to extend a product's durability? What is at
       stake and for whom?   16
   2.4   Example: To reduce or to extend a spacecraft's design
       lifetime?   22

3   **A Brief History of Economic Thought on Durability**   24

   3.1   Introduction: Snapshot from the middle of the story   24
   3.2   Periodization and the history of economic thought on
       durability   26
   3.3   The origins and preanalytic period in the history of economic
       thought on durability: Knut Wicksell and Edward Chamberlin   27
   3.4   Growing interest in durability: Limitations of the price–
       quantity analysis and suspicious industry practices   28

3.5   "Flawed analytic" period in the history of economic
      thought on durability   32
3.6   The Swan-centric period in the history of economic thought
      on durability   33
3.7   The identification of the time inconsistency problem for
      durable goods monopolists   35
3.8   Recent economic literature on durability   41
3.9   Limitations of current economic thinking about durability 44
3.10 Conclusions   48
      Appendix – Origins of Coase's contribution to the time
      inconsistency problem of durable goods monopolists   49

4 Analysis of Marginal Cost of Durability and System Cost
   per Day 53
4.1   Introduction   53
4.2   Nomenclature: Durability, design lifetime, and service life 54
4.3   On values, metrics, and tradeoffs in the search for optimal
      durability   56
4.4   Scaling effects and marginal cost of durability: The example
      of a satellite   61
4.5   Cost elasticity of durability   71
4.6   From marginal cost of durability to cost per day: Regions
      and archetypes   74
4.7   Conclusions   78

5 Flawed Metrics: System Cost per Day and Cost per Payload 81
5.1   Introduction   82
5.2   Two metrics in space system design and their implications 83
5.3   Investigating satellite cost per day   85
5.4   The case for a value-centric mindset in system design   87
5.5   Satellite cost per transponder: Design implications and
      limitations   94
5.6   Conclusions   97

6 Durability Choice and Optimal Design Lifetime for Complex
   Engineering Systems 101
6.1   Introduction: A topic overlooked by economists and
      engineers   101
6.2   An augmented perspective on design and optimization: A
      system's value and the associated flow of service   102

6.3   Optimal durability under steady-state and deterministic
      assumptions   104
6.4   Durability, depreciation, and obsolescence: A preliminary
      account   110
6.5   Uncertainty, risk, and the durability choice problem: A
      preliminary account   118
6.6   Conclusions   123

EPILOGUE. Perspectives in Design: The Deacon's Masterpiece
and Hundred-Year Aircraft, Spacecraft, and Other Complex
Engineering Systems 128

1.   On durability through robustness: The Oliver Wendell
      Holmes way   129
2.   Time to failure   131
3.   Beyond robustness: On durability through flexibility in
      system design   136
4.   The new deacon's masterpiece: Challenge for poets and
      engineers!   141

APPENDIX A. Beyond Cost Models, System Utility or Revenue
Models: Example of a Communications Satellite 145

A.1   Introduction   145
A.2   Motivation: Proliferation of system cost models and
      absence of revenue or utility models   146
A.3   Developing the revenue model structure for a
      communications satellite   149
A.4   Modeling satellite loading dynamics   153
A.5   Integrating satellite loading dynamics with transponder
      lease price   164
A.6   Conclusions   169

APPENDIX B. On Durability and Economic Depreciation 171

B.1   Introduction   171
B.2   Depreciation, deterioration, and obsolescence: The
      traditional interpretation   174
B.3   The model   175
B.4   Depreciation and incremental present value   178
B.5   Depreciation and obsolescence   188
B.6   Concluding remarks   191

Index   195

Preface

Time and the ephemeral nature of human life have been major themes for poets, philosophers, and theologians. Every scripture, philosophical writing, and work of art addresses, explicitly or implicitly, issues of time and the human experience of it.

Engineers have also considered and often grapple with issues of time, except that, instead of the human experience of it, they deal with the relationship of engineering artifacts with time. Less profound than the previous subject but equally thought-provoking is the transiency, not only of human life, but also of human artifacts. Through structural or functional degradation, or loss of economic relevance, the hand of time lies heavy on engineering designs. Several terms are used to describe this particular aspect of a product or a system's relationship with time, namely the duration from fielding a system, that is, when it first enters operation, to its final breakdown or retirement. These terms include *lifespan*, *service life*, *durability*, and *design lifetime*, to name a few. This book discusses these issues in the context of complex engineering systems.

More specifically, this book explores an important issue in engineering design that is becoming increasingly critical for complex engineering systems in general, and aerospace systems in particular, namely the selection and implications of a system's design lifetime. Although economists have grappled with the durability choice problem for simple consumer goods, limited attention has been given to the design lifetime problem(s) of complex engineering systems. The issues at stake in selecting a reduced or an extended design lifetime for an engineering system are complex and multi-disciplinary in nature; they require a thorough engineering understanding of the "technicalities of durability" along with the economic implications of the marginal cost of durability and of the value maximization problem in guiding the durability choice problem.

Systems engineers and program managers recognize the increasing importance of the durability choice problem for engineering systems. For example, design lifetime for infrastructure is typically set at 30–70 years, often with limited rationale, and satellite lifetimes are assigned rather arbitrarily or with limited quantitative analysis (cost-based). This book provides a systemic qualitative and quantitative approach to these problems in the form of a triptych addressing, first, the technicalities of durability; second, the marginal cost of durability, along with the economies of scale (in the time dimension), if any, that result from extended durability; and third, the durability choice problem for complex engineering systems in the face of network externalities (competition and market uncertainty) and obsolescence effects (technology evolution). Because the details of the analyses are system-specific, a satellite example is used in several chapters to illustrate the essence and provide a quantitative application of these analyses.

Also addressed is the increasing tension between the design lifetimes of present-day complex engineering systems and the shortening time scales associated with the obsolescence of their underlying technology base. The book ends with a discussion of the need for and growing interest in the concept of flexibility in system design.

The book is intended for graduate students, researchers, and practitioners. Each chapter is self-contained and can be read independent of the other chapters. The six chapters and Epilogue do, however, tell a coherent story that reaches its climax in Chapters 5 and 6, where traditional engineering wisdom and the "economies of scale" argument in system design are challenged and proved flawed under certain environmental conditions; an alternative framework and solutions are provided in Chapter 6.

Finally, it should be noted that this text is but a small book about a broad topic. It does not pretend to be exhaustive in its treatment of durability related issues. Important topics such as product replacement and recycling, for example, have not been addressed. These topics reflect "downstream" issues that define the end of life, or post-service life, of a system, whereas this book deals with the "upstream" problem of the definition and selection of the intended design lifetime of the system. More specialized texts would do better justice to these subjects of product replacement and recycling than a summary treatment in the present work.

This book is one "panel" of a triptych that consists of two additional books (forthcoming) on flexibility and uncertainty in engineering design. The close connection among time (durability), uncertainty, and flexibility is elaborated in the Epilogue of the present work.

1 Introduction

On Time

1.1 Sundials and Human Time

This story begins with a sundial. One of France's famed fishing ports, at the lower tip of Brittany, is a medieval walled town called Concarneau. A walk though the old city, as in a few others in France, is a spellbinding experience for the visitor. Here, time does not seem to have taken as heavy a toll as in other places. A sundial on the outside wall of one of the old houses in Concarneau carries the inscription *Tempus Fugit Velut Ombra*, which means "time flies [or escapes] like a shadow." I am very fond of this inscription for many reasons. A casual reading of this inscription suggests that it is a clever indication of the obvious: as the day goes by and the Sun traces its path across the sky, the shadow of the gnomon moves along the surface of the sundial ... and time flies, along with this shadow's movement.[1] But the word "like" or *velut* in the inscription, instead of "with" for example, invites the reader to a second, deeper, interpretation: the inscription may be suggesting that, like the shadow, time is elusive and (any presence in it) ephemeral. The inscription, although conveying a sense of fragility, does not decry the destructive side of time, nor does it succumb to the traditional view of time as the destroyer of all things. *Tempus edax rerum*, time the devourer of all things, cried the Roman poet Ovid. This theme found echo in many

---

[1] It is interesting to note that around 400 C.E., when Saint Augustine wrote his *Confessions*, he wrote of "the drops of time" as a metaphor for the water-clock or time measurement device, not the fleeting shadows of sundials.

works of art of later periods. Shakespeare, in the *Rape of Lucrece*, spoke of time as the

> *carrier of grisly care,*
> *Eater of youth [. . .]*
> *Thou nursest all and murder'st all that are:*
> *O, hear me then, injurious, shifting Time!*
> *Be guilty of my death [. . .]*

None of that on our sundial in Concarneau. The power of its inscription, *Tempus Fugit Velut Ombra*, is also in its seemingly unfinished state; like an invitation, it incites the reader to reflect on the consequences of its observation and answer the "so what?" question. How is one to use or spend time given its fleeting nature? The inscription leaves open the possibility of a positive interpretation of time as a provider of opportunities and a "space" for creative endeavors.

That sundial in Concarneau offers a window into much broader and more general questions of time, its meaning, perception, and usage. It also reminds us not to forget the immense influence of time measurement on the history of civilizations.[2] There is a vast literature on the subject. The following pages will touch briefly on some of the issues in order to position the discussion on product durability and system design lifetime in the broader context of human reflections on time in general.

Time and the ephemeral nature of human life have been major themes for philosophers, theologians, and artists. "The human experience of time is all-pervasive, intimate, and immediate" (Fraser, 2003), and, not surprisingly, almost every scripture, philosophical writing,[3] or work of art addresses, explicitly or implicitly, issues of time and the human experience of it. Time

---

[2] Mumford (1967) suggests that "the clock, not the steam engine, is the key-machine of the modern industrial age." Furthermore, Landies (2000) argues that it was time measurement, along with navigational imagination, that "opened the world," and that "without [a common language of time measurement] and without a general access to instruments [of time measurement], urban life and civilizations as we know it would be impossible" (Landes, 2000).

[3] From the Greek philosophers Plato, Aristotle, Plotinus, and others to the 20th century's most influential work by Heidegger, "Being and Time."

remains a deep and thought-provoking mystery.[4] Saint Augustine asks, What is time?[5] in Book XI of his *Confessions*. "Provided that no one asks me, I know; if I want to explain it to an inquirer, I do not know . . . yet what do we speak of, in our familiar everyday conversation, more than of time?" In other words, the knowledge and understanding of (the word) time is instinctive when it is used in a given context, but it becomes difficult when it is looked at in isolation: "The word then becomes an unfathomable enigma" (Miller, 2003).

Physicists realized that the laws of mechanics, in which time is a fundamental coordinate, require a separate implicit assumption of an exogenous flow of time. "We have to assume that there exists a mathematical flow of time," declared Newton. And with this he by-passed the question of the nature of Time. Philosophers, in contrast, posited that time is an experience of the human mind (or soul), which is granted the awareness of time intervals through memory and perhaps some other faculties, and the awareness that the movement of physical bodies in themselves does not constitute time.

But beyond the issues pertaining to its nature lie questions related to the experience of time and the various ways of communicating it. Consider literature, for example: "all literature is about time," writes J. Hillis Miller (2003) in his brief survey of the subject.[6] A number of secondary sources and analyses tend to support Miller's assertion.[7] Similarly, a myriad of human behaviors and creative endeavors finds the original impetus for their existence in an individual's relationship with time. Whitrow (1972), in his account of the nature of time, writes "the mental and emotional tension resulting from man's discovery that every living creature is born and dies,

---

[4] One ancient religion, or more precisely, one branch of the Zoroastrian religion that flourished under the Sasanian empire (circa 226–651 C.E.) was called Zurvanism, from Zurvan, which in Middle Persian (or Pahlavi) means Time. Time, according to Zurvanism, was considered a stronger force than Good (Ahura Mazda) and Evil (Angra Mainyu) and transcended both of them.

[5] The question should not be confused with "what time is it?" More seriously, time measurement raises a set of issues separate from those related to the nature of time. Saint Augustine writes, "I measure time and yet I do not know what I am measuring." Time is not identical with the units by which it is measured.

[6] "The study of time in literature [means] the investigation of the way literary works present in one way or another the human experience of lived time" (Miller, 2003).

[7] For example, George Poulet, "Studies in Human Time," and Paul Ricoeur, "Time and Narrative."

including himself, must have led him intuitively to seek some escape from the relentless flux of time." The classical theme of *ars longa, vita brevis*, which came to mean that art[8] is longer-lasting than the life of the artist,[9] is probably one reflection of an approach to (partially) escaping the flux of time in Whitrow's account. Nietzsche, for example, in *Twilight of the Idols* (1888), illustrates this approach to some extent when he claims that by writing he seeks "to create things on which time tests its teeth in vain; to endeavor to achieve a little immortality in form and in substance – I have never yet been modest enough to demand less of myself." In a similar vein, but unlike Nietzsche, Shakespeare often sought a "little immortality" for his lover through his sonnets. Consider the following, for example:

> *Devouring Time, blunt thou the lion's paws,*
> *And make the earth devour her own sweet brood; [. . .]*
> *O, carve not with thy hours my love's fair brow,*
> *Nor draw no lines there with thine antique pen; . . .*
> *Yet, do thy worst, old Time: despite thy wrong,*
> *My love[r] shall in my verse ever live young.*[10]

In summary, every person has concerned himself or herself with time one way or another, and Landes (2003) is probably right in stating that "all cultures and civilizations have concerned themselves with time, if only to give cues and set bounds to social and religious activities." For the individual, it may be that we are governed by time, just as "we are [physiologically] the children of gravity, which we cannot see or touch, but it has guided the evolutionary destiny of every species, and has dictated the size and shape of our organs and limbs."[11] So perhaps are many of our psychological dispositions,

---

[8] Including artifacts that are not necessarily *artistic* in nature.

[9] Although this is probably a misreading of the original expression ascribed to the Greek physician Hippocrates, in which he probably meant that learning the art of medicine is a long process . . . but life is short.

[10] I also like the following sonnet, which adds another dimension to our present subject:
> *If I could write the beauty of your eyes*
> *And in fresh numbers number all your graces,*
> *The age to come would say "This poet lies: [. . .]"*
> *But were some child of yours alive that time,*
> *You should live twice; in it and in my rhyme.*

[11] D. Newman. "Human Spaceflight from MIR to Mars." AIAA-SF June 2000 Dinner Meeting. Sunnyvale, CA.

behaviors, and actions the children of our experience or relationship with *time* (like gravity, we cannot see it or touch it) and our recognition of the transiency of human life.

1.2 Time and Human Artifacts

Beyond the human relationship with time lie questions related to human artifacts and time. Probably less profound than the discussion in the previous section but equally thought-provoking is the transiency, not only of human life, but also of human handiwork. For example, of all the structures and artifacts of antiquity, only a small number survive today (Terborgh, 1949). Similarly, with regard to more recent artifacts, one often hears about "modern ruins"; abandoned concrete launch pads and steel gantries at Cape Canaveral, remnants of the Mercury, Gemini, and Apollo lunar programs, or the Aerospace Maintenance and Regeneration Center[12] (AMARC) in the Arizona desert, better known as the aircraft graveyard, where over 4,000 aircraft lie moldering in the sun. These modern ruins of engineering systems stand as a reminder that nothing is permanent: through physical or functional degradation, or loss of economic usefulness, the hand of time lies heavy on human artifacts.

Several terms are used to describe this particular aspect of a product or an engineering design relationship with time, namely the span of time from fielding a product to its breakdown or retirement. These include a product's or a system's *lifespan, service life, durability,* or *design lifetime,* to name a few.[13] This book discusses these issues in the context of engineering systems.

1.3 Two Broad Categories of Questions Regarding Durability

Durability is an important multidisciplinary concept. It is traditionally used to describe both an artifact's lastingness, or the extent of its permanence in

---

[12] The AMARC provides storage, regeneration, reclamation, and disposal of aircraft and aircraft parts.

[13] There are differences among these terms, and they will be discussed shortly.

time, and its capability of withstanding use, decay, or wear (*Oxford English Dictionary*). This definition is adequate for the purposes of this introductory chapter. However, in the following chapter, a distinction will be made between the durability of a product in an ex ante and an ex post sense, as well as the design lifetime of a complex engineering system, and more formal definitions of these three concepts will be provided. These details need not concern us for the time being.

For physicians, social scientists, and engineers, a host of critical issues revolves around the notion of durability. Medical doctors and surgeons, for example, are concerned with the durability of living tissue grafts and of prosthetics and implants (a heart valve, for example). Political scientists are interested in the durability of cease-fires, interstate disputes, military regimes, or coalitions in parliamentary democracies. Economists are interested in the choice of durability of durable goods under various market conditions (e.g., monopoly and competition). Engineers are concerned with the durability of structures, concrete, steel bridges, or with the durability of a coating material, of polymer bonds, or of toxic waste after disposal. This is a short list of a few concerns with durability in a variety of settings. Interest in product durability or system design lifetime generally falls under two broad categories of questions, the technicalities of durability and the choice of durability:

1. **The technicalities of durability**: The first category of questions is concerned with the identification and control of parameters affecting durability: What governs durability? What drives the deterioration processes? What limits an artifact's durability, and how can it be made to last longer? For instance, in the case of concrete structures, one can ask: How durable is concrete? How does cold climate affect the durability of concrete? And how can one make concrete more durable, by careful selection of materials, by adjusting mixture characteristics or mixing procedures, or by adding a protective coating? These concerns are generally addressed under the heading of "design for durability" in the civil and structural engineering community. Although the details are discipline-specific, the quest for identifying, understanding, and, to the extent possible, controlling the parameters affecting an artifact's

durability is prevalent in all disciplines (e.g., how to build a more durable bridge, how to make a graft last longer, or how to design a satellite that will remain operational on-orbit longer).

2. **The durability choice**: The second category of questions related to durability is less technical than the previous category and more normative in nature. These questions include the following: How long should artifacts be made to last? What is the purpose of an artifact's durability and what metrics are to be optimized – maximized or minimized – through this choice? In the case of consumer goods, for example, how do manufacturers select product durability under various market conditions (e.g., monopoly and competition)? Or, in the case of industrial goods, for example, what should customers ask or require that the manufacturer or contractor provide as a system's durability?

Questions related to the choice of durability have received significant attention from economists, but little attention has been given to these questions from the technical or engineering community. This limited interest may in part be due to the fact that engineers often view durability as a constraint rather than as a choice. Engineers often strive to create products with the longest durability technically or practically achievable. Engineering efforts in various cases can often be interpreted as (1) pushing the boundary of the technically achievable durability of a product or a design or (2) reducing the cost at which the current durability is achieved. In other words, engineers to date have been more interested in the technicalities of durability questions.

There are important and subtle durability choice questions that are by no means fully addressed yet. For example, because "infinitely durable" components or systems do not exist, durability specification requires choices and tradeoffs. System engineers and program managers, in deciding how much durability is needed, must assess how much durability is worth and how much customers are willing to pay for it. This theme will be further developed throughout this work. But although the technicalities of durability have received significant attention in the engineering literature, limited attention has been given to the durability choice problem in engineering systems. Systems engineers and program managers are beginning to recognize the importance of the durability choice problem for engineering systems. Consider the following observation from a civil engineer:

> Design service lives for infrastructures are set typically at 30–70 years often with very limited rationale. Definition of design service life [is], in principle, a choice to be made by designers and owners, based on life-cycle costs and benefits. Most typically, no such analysis is conducted. (Lemer, 1996, p. 155)

A similar mindset can be found in the space industry:

> In principle, we would like to obtain a graph of a spacecraft cost versus design lifetime.... In practice, [such analyses] are almost never done or at best, are done qualitatively. The mission duration is normally assigned rather arbitrarily.... Thus, there may be a push to produce spacecraft lasting five or ten years because people believe these will be more economical than ones lasting only a few years. Doing [these analyses] provides a much stronger basis for determining whether we should push harder for longer spacecraft lifetime or back off on this requirement. (Wertz and Larson, 1999, pp. 17–18)

The purpose of this book is to identify and discuss what some of "these analyses" cited by Wertz and Larson are or should be and to contribute an analytical framework toward a rational choice of durability for engineering systems, from a customer's perspective, and in the face of network externalities and obsolescence effects.

1.4 Why the Interest in Product Durability and System Design Lifetime?

It is likely that questions of durability were raised and became of interest to academics as soon as the notion of a durable good was conceptualized. A *durable good* is an economic term that designates a set of products that provide utility or a flow of service over a period of time, as opposed to products that are immediately consumed on first use. Consider the two ends of the product durability spectrum. At one end are goods that never wear out – sometimes referred to as perfectly durable goods. At the other end, we have nondurable goods that are totally consumed when used once. Real world durable goods occupy the space between these two ends of the durability spectrum. Questions of durability, both technical and economic, are bound to arise, as soon as one realizes that durable goods are durable to some extent: To what extent are they or should they be made durable? How is the durability choice made, and what needs to be taken into account in making this choice? How does it impact a manufacturer's profits? And subsequently, how does an industry structure impact the choice of durability?

There is a popular belief that manufacturers of durable goods often deliberately reduce the durability of their products to increase sales and profits. There are interesting case histories in the electric lamp, razor blade, and vacuum tube industries that suggest that producers of these durable goods may have colluded to limit the durability of their products, or had a concealed policy of deliberately limiting their products' life, in order to increase their sales "when customers' interests were generally thought to be better served by [products] of much longer life" (Avinger, 1968).

The discussion in the previous paragraph is meant to serve two purposes. First, it introduces three main stakeholders who should be taken into account in analyzing issues of product durability and system design lifetime. These are (1) the customers, (2) the manufacturer, and (3) society at large. Second, the previous paragraph indicates a tension between the stakeholders above, as each is affected differently by an extended or reduced product lifetime, and shows that the interests of one are not necessarily aligned with the interests of the others. One should therefore recognize that, in exploring the issues at stake in reducing or extending a product durability, it is necessary to first specify from which stakeholder's perspective the analysis is carried out, as the interests and tradeoffs can be substantially different.

Durability became a contentious issue,[14] and the practices alluded to above heightened the interest of academics (mainly economists) in durability choice under various market structures. They asked: Will durability choice be different under different conditions of competition and monopoly? Empirical or anecdotal evidence seemed to suggest that monopolists would indeed produce goods of shorter durability than competitive markets. Starting in the 1960s, the economic literature saw a proliferation of studies on durability.

---

[14] This is further discussed in Chapter 3.

In summary, both the engineering and economic issues associated with product durability and system design lifetime continue to offer a rich field of investigation for academics and industry professionals. The implications for selecting a reduced or extended product durability are complex and multidisciplinary in nature; they demand careful consideration and require much more attention than they have received to date in the engineering and economic literature, as their impact is substantial and can ripple throughout an entire industry value chain.

1.5 Book Organization

This book, as mentioned previously, deals with issues of durability and system design lifetime, with a focus on engineering systems rather than consumer goods. Although the arguments in each chapter build on those of the previous chapters,[15] they are nevertheless designed to be self-contained. Although this makes it easy for the reader to read and understand any one chapter from the book without reading the previous chapters, it implies that there is a bit of overlap between chapters. The reader who wishes to go through the whole book in one sitting can easily skip through the overlapping parts.

Chapter 2 explores the qualitative implications associated with reducing versus extending a product's durability or a system design lifetime, as seen from the perspective of the customer, the manufacturer, and society at large. This chapter shows that the implications of selecting a reduced or an extended product durability are complex and multidisciplinary and affect an entire industry value chain. One should not reduce the subject, as is often done in the economic literature, to the study of revenues or profits for the manufacturers.

Chapter 3 provides a narrative of development of economic thought on durability. This chapter also discusses the present limitations in the economic literature on durability and provides a background against which the remaining chapters can be contrasted.

---

[15] With the exception of Chapter 3, which is an overview of economic thought on durability.

Chapter 4 introduces the concept of marginal cost of durability and shows why it is a prerequisite for addressing the durability choice problem of engineering systems. Because the details of a marginal cost of durability analysis are system-specific, this chapter uses a satellite system example to illustrate the essence and provide a quantitative application of such analysis. This chapter also introduces two related metrics, the cost elasticity of durability and the cost per day of a system. These metrics are designed to elucidate the economies of scale (in the time dimension), if any, that result from extended durability. In short, this chapter contributes a necessary first step toward a rational choice of durability for engineering systems from a customer's perspective.

Chapter 5 challenges the traditional "economies of scale" argument in engineering design with respect to both durability and system capability. Decision-makers often invoke the "economies of scale" argument to justify the design of larger, more capable, and longer-lived engineering systems. This chapter makes the case that two metrics often used to guide design choices in the space industry, and traditionally associated with the economies of scale argument, namely *cost per day* and *cost per payload* (or capability), are flawed under certain environmental conditions and result in design choices – increasingly longer-lived systems and larger payloads – that do not necessarily maximize the system's value. This chapter also advocates shifting the emphasis from cost to value analyses to guide design choices and makes the case that dynamic environmental conditions or competitive markets require a value-centric mindset that views an engineering system as a value-delivery artifact and integrates considerations about the system's cost, its technical environment, and the environment it is serving in order to make appropriate system design choices.

Chapter 6 addresses the durability choice problem, as seen from the customer's perspective, and in the face of network externalities and obsolescence effects. In this chapter, analytical results for optimal durability are derived under steady-state and deterministic assumptions, and trends or functional dependence of the optimal durability on various parameters are identified and discussed. In addition, this chapter explores the durability choice problem when the risk of obsolescence is accounted for; it also investigates the durability choice problem under uncertainty and discusses the various risks in making *cautious* or *risky* choices of durability. In short, Chapter 6 provides an analytical framework for a (customer's) rational choice of durability, and the analyses here provided ought to be made available to decision-makers to support in part the durability specification of engineering systems.

Finally, it should be noted that this text is but a short introduction to a broad topic. It does not pretend to be exhaustive in its treatment of issues related to the durability of engineering systems. Important topics such as product replacement or recycling, for example, have not been addressed. There are specialized texts on these subjects for the reader wishing to explore these topics in greater depth.

The *Epilogue* to this text is intended as an entertaining but thought-provoking look at one particular problem related to durability that has not been addressed in the previous chapters, but that is becoming ever more important: the increasing tension between the design lifetimes of present-day complex engineering systems and the various time scales associated with the obsolescence of their underlying technology base. The Epilogue ends with a brief discussion of the need for and growing interest in the concept of flexibility in system design.

REFERENCES

Avinger, R. L., Jr. "The Economics of Durability." Ph.D. dissertation, Duke University, 1968.

Fraser, J. T. "On ye shoulders of giants" in: *What is Time?* New Introduction to Whitrow (1972), 2003.

Landes, D. S. *Revolution in Time.* Harvard University Press, Cambridge, MA, 2000.

Landes, D. S. "Clocks and the wealth of nations." *Dædalus,* Spring 2003, pp. 20–6.

Lemer, A. C. "Infrastructure obsolescence and design service life." *Journal of Infrastructure Systems*, 1996, 2 (4), pp. 153–61.

Miller, J. H. "Time in literature." *Dædalus,* Spring 2003, pp. 86–97.

Mumford, L. *The Myth of the Machine: Technics and Human Development.* Harcourt, New York, 1967.

Newman, D. "Human spaceflight from MIT to Mars." AIAA-SF, June 2000 Dinner Meeting, Sunnyvale, CA.

Terborgh, G. *Dynamic Equipment Policy.* McGraw-Hill, New York, 1949.

Wertz, R., and Larson, W. *Space Mission Analysis and Design*, 3rd ed. jointly published by Microcosm Press, Torrence, CA, and Kluwer Academic Publishing, Dordrecht, Boston, London, 1999.

Whitrow G. J. *What Is Time? The Classic Account of the Nature of Time.* Oxford University Press, Oxford, 1972.

2 To Reduce or to Extend Durability?

A Qualitative Discussion of Issues at Stake

PREVIEW AND GUIDE TO THE CHAPTER

As we discussed in the first chapter, the implications for selecting a reduced or extended product durability are complex and multidisciplinary in nature. They affect an entire industry value chain and should not be reduced solely to the revenues or profits for the manufacturers resulting from the sales of the goods (as is often done in the economic literature on the subject). There are tangible and intangible consequences from the decision to reduce or extend a product's durability or a system's design lifetime. This chapter discusses the qualitative implications associated with the choice of reducing or extending a product's durability or a system design lifetime, as seen from the perspective of three stakeholders, the customer, the manufacturer, and society at large.

2.1 Introduction

As discussed in the first chapter, there is a popular belief that probably dates back to the 1930s that manufacturers of durable goods often deliberately reduced the durability of their products in order to increase their sales and profits. There is some historical evidence that gives credence to such a belief in the particular cases of the electric light bulb, razor blade, and vacuum tube industries. However, it is generally agreed that consumers would have been better served by longer-lived products.

Consider, for example, the electric light bulb industry. This industry was highly cartelized in the first half of the 20th-century,[1] and its cartel often took

---

[1] See, for example, the discussion in Stocking and Watkins (1946).

proactive measures to limit lamp life or the durability of the bulb, with the obvious objective of increasing sales for the cartel members.[2]

Several industries, however, strongly denied having a concealed policy of either deliberately limiting product operational life or accelerating product obsolescence by introducing upgrades or new functionalities into a product in order to promote customer dissatisfaction and promote sales of new products.

Limited product durability, not just of light bulbs but of more bulky goods, such as car batteries, sparked environmental concerns among ecologists and policy makers and created interest in the contribution that extended product design lifetime can make toward reducing waste and other environmental problems.[3]

The brief discussion in the previous paragraphs is meant to serve a specific objective: to illustrate the fact that there are different stakeholders that need to be reckoned with when analyzing issues of durability, namely the customer, the manufacturer, and society at large, and that the interests of these stakeholders are not necessarily aligned with respect to product durability, as they are affected differently by reduced or extended durability.

The next sections introduce nomenclature adopted in this text and provide a qualitative discussion of the implications associated with a product's durability or system design lifetime, as seen from the perspectives of our three stakeholders.

2.2 Nomenclature: Durability and Design Lifetime – A Matter of Connotation

Product durability and system design lifetime are similar in that they both characterize an *artifact's* relationship with *time*. The two expressions, however, sometimes carry different connotations that deserve to be recognized. *Product durability* has often been used to characterize the durability

---

[2] See Robert Avinger's article "Product Durability and Market Structure: Some Evidence." *Journal of Industrial Economics*, 1981, *29*(4), pp. 357–74.

[3] See, for example, the O.E.C.D. report entitled "Product Durability and Product Life Extension: Their Contribution to Solid Waste Management." Organization for Economic Co-operation and Development, Paris, 1982.

of a consumer good. *Consumer goods* is used here to describe products of limited complexity. In contrast, *design lifetime* is often associated with complex engineering systems or capital goods. No such distinction is made in the following pages, and *durability* and *design lifetime* are used interchangeably. Furthermore, unless stated otherwise, *durability* is used in an ex ante sense; that is, it designates the requirement that the manufacturer selects for the *intended duration of operation of the system* – alternatively, this requirement can be imposed on the manufacturer by the market, customers, or regulators.[4] Designers and engineers will make different design choices for a system's structure and components depending on whether the selected design lifetime is, for example, one year, five years, or ten years.

2.3 To Reduce or to Extend a Product's Durability? What Is at Stake and for Whom?

As mentioned in the first chapter, the implications for selecting reduced or extended durability are complex and multidisciplinary in nature; they affect an entire industry value chain and should not be reduced solely to a discussion of the manufacturer's revenues. There are in fact tangible and intangible consequences of reducing or extending a product's durability or a system's design lifetime, as will be shown shortly.

Let us look at some qualitative implications associated with reducing versus extending a product's durability or a system design lifetime, as summarized in Table 2.1.

Each entry in the table is tagged with a numeral followed by an "A" or a "D" for what appears more as an advantage or a disadvantage from the perspective of each stakeholder. It should be noted that in some cases the

[4] Who, in an industry value chain, makes the decision for a product durability or system design lifetime reflects to some extent where market power resides. For example, in the satellite business (or value chain), satellite manufacturers rarely decide on a spacecraft design lifetime. It is primarily the customer, the satellite operator, who imposes the spacecraft design lifetime on manufacturers. This present situation is a reflection of both the industry structure and the overcapacity in satellite production compared with levels of demand.

"customer" of a system is not the end-user.[5] In general, there are numerous stakeholders for complex engineering systems,[6] and they are all affected to some extent, and often differently, by the system's design lifetime. Table 2.1 is therefore not meant to be exhaustive but is confined to the three stakeholders mentioned above. The following subsections expand on the information in Table 2.1.

2.3.1 Implications of Reduced Durability

From a customer's perspective, a product or a family of products with a shortened lifetime is more likely to be improved on, during a given time period, through more frequent iterations of fielding (or market rollout of the product) and feedback to the manufacturer, when compared with products with longer lifetimes. Alternatively, customers may perceive the need to purchase increasingly many products as their lifetimes decrease negatively if the duration of the needed service exceeds the system design lifetime.

From a manufacturer's perspective, the two points raised above translate into several advantages. First, manufacturers of products or a family of products with shortened lifetimes have an increased ability to improve their products through more frequent iterations of fielding and customer feedback. Second, shorter lifetimes can stimulate sales, because customers need to buy more product to sustain the same level of service during a given time period. Another implication of shortened lifetimes is a heightened need for manufacturers to remain technically up to date and attentive to the voice of the customer in order to fend off competitors. Because customers of systems with short design lifetimes are not locked in (compared with customers who acquire longer-lived products), they can recommit resources to acquiring new products or systems from the competition if the incumbent is not constantly offering best-value products. Manufacturers who reduce their systems' design lifetimes have limited opportunity to generate revenues

---

[5] For example, satellite operators are "customers" of satellite manufacturers; however, the end-users are TV broadcasters, communications carriers, service providers, or others.

[6] In the satellite industry example, stakeholders other than the satellite manufacturers and operators include end-users, launch services, equipment manufacturers, insurance companies, and regulatory agencies.

Table 2.1. *Stakeholder's perspective for a product's durability*

	Reduced durability			Extended durability		
	Customer's perspective	Manufacturer's perspective	Society's perspective	Customer's perspective	Manufacturer's perspective	Society's perspective
	1A. Products more likely to be improved through more frequent iterations of fielding and feedback to the manufacturer than products with longer lifetimes.	1A. Ability to improve subsequent products through more frequent iterations of fielding and customer feedback.	1A. Shorter design lifetime can stimulate innovation and technological progress.	1A. Smaller volume of purchases.	1A. Greater durability allows manufacturers to create additional product support products such as service contracts that generate additional revenue in addition to the sale of the product or system.	1A. Products with longer design lifetimes result in less waste during a given time period than those with shorter lifetimes.
		2A. Potential for higher unit sales volume over time.	2A. Potential for maintaining and boosting industry employment level through higher sales volume.	2A. Potentially smaller cost per operational day.	2A. Increased design lifetime acts as a magnifier of "reliability as a competitive advantage." Product reliability is less critical for products with short lifetimes than those with longer lifetimes.	2A. Longer design lifetime can stimulate the creation of a secondary market for the products.

1D. Need to purchase more products for a given duration.	1D. Fewer opportunities for revenues from services.		
	1D. Adverse environmental effect as a result of more product disposal during a given time period.	3A. "Old products" easier to replace than repair – hence the likelihood of more state-of-the-art products in use than with products with longer lifetimes.	3A. Heightened obligation for employees to remain technically up to date and attentive to the voice of the customer.
	1D. Increased risk that the product will be technically or commercially obsolete before the end of its lifetime – hence loss of revenues.	1D. Extended warranty needed, which may result in higher levels of unpaid services.	3A. Increased durability as a possible strategy to deter entry of potential competitors (increased barrier to entry).
			1D. Increased risk of technological slowdown, potential increase in an industry's unemployment.

This table was developed in collaboration with Juan Palo Torres-Padilla. His contributions are gratefully acknowledged.

from service contracts. This opportunity loss should be carefully compared with the increased volume of sales and revenues associated with shorter lifetimes before manufacturers decide whether they are better off reducing or extending their products' durability.

From society's perspective, short design lifetimes may present several advantages. First, shorter design lifetimes can stimulate innovation and technological progress. Planned obsolescence, which can be loosely defined as the intentional reduction of product durability coupled with rapid innovation, may be preferred, from society's perspective, to long-lasting products and limited innovation. Second, if the assumption discussed above is true, then this increased sales volume has the potential to maintain or boost industry sales and employment. Third, as a product ages, say, for example, a product that has been in service for three years, it is often easier to replace, should it fail, than to repair if it has a lifetime of four years rather than, say, six years. This observation implies that, at any given time, more state-of-the-art products are likely to be found in use than if these products were designed with longer lifetimes. One adverse environmental effect associated with shortened lifetimes is increased waste during a given time period when compared with longer-lived products.

2.3.2 Implications of Extended Durability

From a customer's perspective, purchasing products or systems with long lifetimes offers two main advantages.[7] First, customers have to purchase fewer products or systems for the duration of their service needs as design lifetime increases. Second, it is more likely that the product's or system's cost per operational day decreases as the system's design lifetime increases. This point will be discussed in more detail in the following analytical sections. One disadvantage customers will encounter with longer-lived systems is an increased risk that these systems will be technically and commercially obsolete before the end of their lifetimes, and hence have an increased risk of loss of revenue.

---

[7] The reader will notice that some of the stakeholders' advantages in reducing system design lifetimes transform into disadvantages when longer design lifetimes are considered, and vice versa.

From a manufacturer's perspective, there are three main implications associated with an increased system design lifetime. First, systems with longer design lifetimes offer an increased ability to generate additional revenue and profit from service contracts rather than from the sale of the system (it is worth noting that, for satellites, manufacturers normally do not have service contracts, but usually provide anomaly support through the contracted life on-orbit at no cost to the operator). There is limited potential for additional revenue from service contracts with a system of short design lifetime. The second implication, which is neither an advantage nor a disadvantage, merely an observation, is that increased design lifetime acts as a magnifier of a system's "reliability[8] as a competitive advantage." That is, the reliability of a system is increasingly more valuable for customers as the system design lifetime increases. Therefore, manufacturers with core competencies or the know-how to produce highly reliable systems have the competitive or sales incentive to increase their system design lifetimes in order to augment the quality gap with manufacturers of less reliable systems. Manufacturers who extend their system design lifetimes may have to bear the additional costs in unpaid services of extended warranties. This risk is heightened for manufacturers of less reliable systems. In other words, manufacturers who do not have a track record of designing reliable systems should carefully consider before engaging in "design lifetime extension behavior" to differentiate their systems from the competition's. This risk should be weighed against, or can be mitigated by, the service contract advantage discussed above. The third implication in choosing extended durability is that manufacturers may choose increased durability as a strategic deterrent or barrier to entry of potential competitors.

From society's perspective, one clear environmental advantage of systems with long design lifetimes over short-lived products is that they result in less waste during a given time period. Another implication is that long design lifetimes can stimulate the creation of a secondary market for products, and

---

[8] Reliability is a popular concept that has been celebrated for years as a commendable attribute of a person or a product, as being trustworthy or sure, and in which confidence may be put (as to its operation in the case of product). In an engineering context, reliability is formally defined as the probability that a component or a system will perform its intended function, under given environmental conditions, for a specified period of time.

hence increased economic activity. One disadvantage of fielding systems with increasingly long lifetimes may be that, although short design lifetimes can stimulate faster innovation, long design lifetimes can increase the risk of technological slowdown and adversely impact an industry's employment level.

2.4 Example: To Reduce or to Extend a Spacecraft's Design Lifetime?

Over the past two decades, telecommunications satellites have seen their design lifetimes on average increase from 7 to 15 years. Increased spacecraft lifetime came about because of technological innovation. It became technically possible to design satellites with longer lifetimes with better battery technology and more reliable transponders. Additionally, satellite operators determined that, given the market conditions, longer-lived satellites would increase their return on investment. In other words, there was a demand for increased durability, and technological innovation made it possible to meet this demand. Satellite manufacturers also determined that longer-lived satellites are good advertising as a testament to the quality of their products and corresponded to what their customers were asking for. A more detailed analysis is necessary to determine whether this behavior, competing on durability, is hurting the satellite manufacturing industry by limiting the need for additional satellites and shrinking the replacement market.

Extending satellite design lifetimes has several side effects. On one hand, the manufacturers must produce larger and heavier satellites as a result of several factors, such as additional propellant for orbit- and station-keeping and increased power generation and storage capability. This in turn increases the satellite's development and production cost. On the other hand, as the design lifetime increases, the risk that the satellite will become obsolete, technically and commercially, before the end of its lifetime increases. This tradeoff is illustrated in Figure 2.1.

The discussion above indicates that, in specifying a spacecraft design lifetime requirement, operators have to assess the risk of loss of value due to both obsolescence of their spacecraft technology base and the likelihood of changing or shifting market needs after the satellite has been launched. For

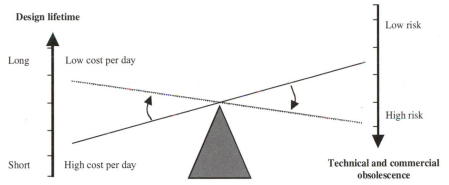

Figure 2.1. Design lifetime tradeoffs: keeping a satellite cost per operational day low through long design lifetime but risking that the satellite will become obsolete before the end of its lifetime.

example, it is not clear that it is in the best interest of a satellite operator to make the contract life of a spacecraft too long: new or enhanced capabilities, such as better spatial resolution for an optical instrument, might be developed and become available within a couple of years following the launch, and hence a need might arise to launch a new satellite or risk losing market share to a competitor who launches later with newer or more advanced capabilities.

So how can one capture these various effects and their impact on the value of a system? The following chapters address these issues, starting with a brief history of economic thought on durability in the next chapter.

REFERENCES

Avinger, R. L., Jr. "Product durability and market structure: Some evidence." *Journal of Industrial Economics*, 1981, 29 (4), pp. 357–74.

Butlin, J. A. "Product Durability and Product Life Extension: Their Contribution to Solid Waste Management." Organization for Economic Co-operation and Development, Paris, 1982.

Stocking, G. W., and Watkins M. W. *Cartels in Action*. The Twentieth Century Fund Press, New York, 1946.

3 A Brief History of Economic Thought on Durability

PREVIEW AND GUIDE TO THE CHAPTER

This chapter provides an overview of economic thought on durability. The history of the dynamics and evolution of the durability choice problem is broken into four periods: (1) the origins and preanalytic period, (2) the flawed analytic period, (3) the Swan-centric period, and (4) the identification of the time inconsistency problem for durable goods monopolists. The narrative thread shows interesting cycles and a pattern of rise and then stumbling, followed by a more subtle and careful development of an economic theory of durability. The chapter concludes with a discussion of the present limitations in this literature.

3.1 Introduction: Snapshot From the Middle of the Story

In a review article on the effects of market structure on the choice of product durability, Richard Schmalensee (1979) concludes his survey by discouraging further research into this area and suggests that "the study of durability may have encountered strongly diminishing returns."

Despite his warning, many scholars continued to investigate issues of durability, and major contributions were made in this area in the next two decades by economists including Jeremy Bulow (1982, 1986), Nancy Stokey (1981, 1988), and Michael Waldman (1993, 1996), to name a few.

Schmalensee's comments were in part justified because economists had focused too much, for too long, and almost exclusively on the assumptions underlying Peter Swan's counterintuitive result that product durability is invariant to market structure, or, more simply, that manufacturers of durable goods make the same choice of product durability irrespective of whether they are under competitive market conditions or monopoly (Swan, 1970, 1971, 1972; Sieper and Swan, 1973).

Why this fixation on Swan's result and assumptions? There are two related reasons that can explain this emphasis. The first reason is that in the decade prior to Swan's publications, several economists such as Martin (1962), Kleiman and Ophir (1966), Levhari and Srinivasan (1969), and Schmalensee (1970) himself had argued that a monopolist would produce goods of less or shorter durability than a manufacturer in a competitive market. It was believed that monopolists underinvest in durability, whereas consumers would be better served by longer-lived products. This was an accepted, noncontentious result. Swan's result challenged this traditional wisdom and exposed the flaws in the previous authors' analyses. Swan stated that "the monopoly and competitive solution to the durability question [are the same]: firms will select the same degree of durability regardless of monopoly power" (Swan, 1972). In his words: "the assumptions on which their [previous authors', cited above] analyses are based are, at least in part, contradictory and therefore inappropriate for the analysis of durability" (Swan, 1970). This conclusion stirred the economic community interested in this subject. Seven assumptions underlie Swan's counterintuitive result, and many economists focused for the remainder of the decade (1970s) either on questioning these assumptions or on testing the robustness of Swan's result to relaxing some of these assumptions (Schmalensee, 1979; Waldman, 2003). The second related reason that can in part explain the fixation on Swan's findings in the 1970s was that his results appeared formally sound, significantly counterintuitive, and profoundly unsatisfactory.

In summary, these two reasons, (1) that many economists (and generally accepted results) have been proven wrong with respect to durability choice and monopoly and (2) that the new proposed results seemed technically sound but highly unsatisfactory, resulted in the overemphasis on Swan's findings and underlying assumptions in the 1970s. By the end of the decade, however, limited progress had been made and "not surprisingly, Swan's results [were found to be] robust to relaxing some assumptions, but not others" (Waldman, 2003). Research has continued to focus on the same durability and market structure question, which was becoming of limited appeal and relevance (as will be discussed shortly), and using approaches similar to the ones already in use, which were unlikely to provide any new insights. It is in this context that Schmalensee's warning can best be understood, that "the study of durability may have encountered strongly diminishing returns" (Schmalensee, 1979): one decade of focus on Swan's result was enough, and it was time to move on.

The contributions that were made in the general area of product durability, and its corollary, planned obsolescence, in the 1980s and 1990s resulted from thinking anew about these issues by authors who were unencumbered by Swan's models. These new contributions and insights were in fact based to a large extent on a conjecture first proposed by Coase (1972), as will be discussed in the following section on the history of durability research. In short, new contributions were made because researchers had not confined themselves to the proverbial box (outside which people are often advised to think) that Swan's models and findings had provided.

But despite the progress made in the post-Swan era, major limitations still remain unaddressed in the study of durability, and the proverbial box may in fact be better described in this case as nested boxes – like Russian matryoshky or stacking dolls, studies of product durability extricated themselves from one box only to find themselves confined to another. These limitations will be discussed after the following review of the history of the rise, stumbling, and development of economic thought on durability.

3.2 Periodization and the History of Economic Thought on Durability

This section provides an overview of the history and evolution of economic thought on durability. The history of durability can be conceptually divided into four periods:

- The origins and preanalytic period (prior to the 1960s)
- The "flawed analytic" period (the 1960s)
- The Swan-centric period (the 1970s)
- The identification of the time inconsistency problem of durable goods monopolists: Coase (1972) and the post-Coase period (the 1980s to the present).

Historians often engage in such periodization exercises. The exercise can be useful in helping to conceptualize and communicate certain dynamics, but *periods* should not be taken too literally, as they are rarely linear, they often overlap, and their boundaries are seldom clearly identifiable.

3.3 The Origins and Preanalytic Period in the History of Economic Thought on Durability: Knut Wicksell and Edward Chamberlin

It is likely that questions of durability were raised and became of interest to academics as soon as the notion of a durable good was conceptualized.

The Swedish economist Knut Wicksell (1851–1926) is traditionally associated with the origins of economic thought on durability. In his *Lectures on Political Economy* (translated 1934; the original text in Swedish was written in 1901 and 1923), Wicksell notes that "a durable good is given durability in order that it should yield more services, but these must necessarily be postponed to a more or less remote future." He proposes to treat the "lifetime of capital-goods as if it were altogether separated from their other property" and suggests that "machines [. . .] be constructed to last long enough to be remunerative." The (flawed) idea that consumers are always better served by longer durability may date back to Wicksell when he states, for example, "a farmer has to choose between two ploughs, one which lasts ten years, and the other lasting eleven. If he chooses the more durable and dearer plough, he has the benefit of an extra year's service," or more explicitly, "[it is] always advantageous to increase the durability of the machine."

Following Wicksell, Edward Chamberlin (1899–1967), the Harvard economist who coined the term "product differentiation," is the person who truly set the agenda for research on durability issues:

> Durability is an aspect of products which is exceedingly variable. . . . Since it is variable, the producer has to face the question of how durable to make his product. Evidently if he makes it too durable, as soon as people have bought one unit they will not need another for a substantial period during which there will be no repeat demand for his product. He has an interest then in making it less durable. . . . On the other hand, just as he must not set his price too high, so he must not offer a product [that] wears out too fast in comparison with others on the market. The problem is to find the length of life of his product which will maximize his profits. (Chamberlin, 1953, pp. 23–24)

Chamberlin's writings, more nuanced than Wicksell's, suggest that "in some cases, as when there is a style factor, buyers do not want durability beyond a certain point." In other words, buyers are not necessarily always happier requesting products with increasingly longer durability in contrast to Wicksell's position. The research agenda on product durability and market structure that has occupied economists for the past 40 years was first formulated by Chamberlin's 1953 article, in which he stated,

> The question is raised: what governs durability of different products, and under different conditions of competition and monopoly; and how does the optimum defined by profit maximization compare with standards defined by the public interest or welfare criterion? (p. 24)

It is surprising and unfortunate that Chamberlin's work is rarely, if ever cited[1] in subsequent studies, literature reviews, or surveys of durability. It is fair, however, to acknowledge him as the person who first identified this vast area of academic investigation (durability and market structure), even though he did not publish any analytical work on this subject.

3.4 Growing Interest in Durability: Limitations of the Price–Quantity Analysis and Suspicious Industry Practices

Before we proceed to the second period in the history of economic thought on durability, it is appropriate to point out two facts that contributed to the growing academic interest in durability. The first is an increasing recognition of the limitations of the traditional price–quantity analysis that pervaded economic thinking, and the second is the identification of suspicious industry practices (monopolists and cartels) with respect to product durability.

3.4.1 Limitations of the Traditional Price–Quantity Economic Analysis

In 1979, Schmalensee lamented that "very little formal analysis of product quality determination was to be found in the literature." He was in fact echoing a concern articulated by Chamberlin in 1953, as will be discussed

---

[1] With the exception of Martin (1962).

shortly. Schmalensee suggests that one "obvious reason for this persistent neglect is the apparent difficulty of mathematics to model decision-making in this dimension. While price and quantity can naturally be treated as scalars, it is far from obvious that any single mathematical representation of quality can serve for a broad spectrum of products" (Schmalensee, 1979).

It should be noted that *quality* in early economic writings (prior to the 1980s) referred to product attributes in general and did not carry the strong reliability connotation that is associated with it today.

Chamberlin, in his 1953 paper[2] titled "The Product as an Economic Variable," was perhaps one of the first economists to suggest the inadequacy of price–quantity analysis and the need to account for product attributes, or quality, in economic analyses:

> It is not enough to concern oneself with the proposition that the demand for a given product varies with its price; one must also take into account that the demand at a given price varies with its quality.... Products are not in fact 'given'; they are continuously changed – improved, deteriorated, or just made different – as an essential part of the market process.... What an absurdly small part of a very broad problem is the first pair of variables price–quantity relationships, to which economic theory has traditionally limited itself. (p. 3)

Durability was (and still is in many cases) considered to be one aspect of a product's quality.[3] This product attribute, durability, offered one of the easier ways to analytically expand the traditional price–quantity economic analysis into a third dimension of product quality. For example, Kleiman and Ophir (1966) described the rationale for their analysis of durability in the following terms:

> Price theory is essentially a two-dimensional analysis. It relates quantities to cost or price, abstracting from the existence of other attributes of the product, such as quality. This paper deals with a particular case of a third dimension, that of durability. (p. 165)

---

[2] A draft of which, he claims, existed as far back as 1936.

[3] See for an alternative view Fishman et al. (1993). I believe, however, that these authors confused the various evolving connotations of "quality" and proposed to separate durability from the recently acquired reliability connotation of quality.

Growing academic interest arose in issues of durability in the second half of the 20th century because economists recognized the limitations of the traditional price–quantity analysis and sought to expand it by including an additional dimension of product quality. Durability provided one dimension of product quality that could be modeled mathematically without much difficulty. Significant progress was made since the times of Wicksell and Chamberlin, and by the end of the 1970s, Schmalensee (1979) could claim that "the best developed subset of the [economic] literature [that incorporates product quality] treats the quality measure as some function of the durability of a good, [that is], the potential products differ mainly in their durability."

3.4.2 Suspicious Industry Practices

Another factor that contributed to the growing academic interest in durability may be that durability became a contentious issue. How did this happen?

As mentioned previously, there is some historical evidence that suggests producers of durable goods may have colluded to limit the durability of their products. For example, the electric lamp industry in the first half of the twentieth century seems to have been highly cartelized (Stocking and Watkins, 1946). In the United States, it was dominated by General Electric, which owned the basic patents and controlled licensing arrangements with other producers (Avinger, 1981):

> In 1924, under GE's direction, the producers of light bulbs formed a worldwide cartel. Through a Swiss corporation, Phoebus, producers agreed to divide markets, share patent know-how, and dictate quality standards including lamp life. Although the cartel claimed that these standards were aimed at maintaining high lamp quality, such standards also served to limit lamp life. (p. 359)

An internal memorandum from Phoebus seems to confirm Avinger's observation:

> All manufacturers are committed to our program of standardization, as well as our formula for arriving at the economic life of lamp. This is expected to double the business of all parties within five years. (Stockings and Watkins, 1946, p. 354)

A good discussion of the history of the electric lamp industry can be found in Bright (1949).

Similar practices have been suspected in the razor blade industry.[4] Gillette, for example, which in 1962 accounted for 70% of the market share for razor blades, was strongly committed to its short-lived carbon-steel blade. But already in 1953, Chamberlin, who taught in Cambridge, Massachusetts, close to the Gillette headquarters, stated: "I have been told by people who ought to know, that it is possible to make at reasonable cost a razor blade which would last a lifetime.... What is clear is that if razor blades of this kind were made, as soon as everybody had one there would be no more sold." Gillette's market domination lasted until 1962, when Wilkinson introduced its long-lasting stainless-steel blade. There was apparently nothing new about that blade when Wilkinson introduced it – Gillette even claimed to hold patents on the products – but "all the major [producers] had not pushed it for obvious reasons: since a stainless-steel blade lasts longer than a carbon-steel blade, they would obviously sell fewer of them" (Avinger, 1981). Even Gillette's CEO at the time stated, referring to the stainless-steel product, "when other people come in, we have to join. But we wouldn't have chosen this route" (Avinger, 1981). In other words, Gillette delayed the introduction of a longer-lasting product because it feared it would negatively impact its sales and profits. A monopolist, or more fittingly in Gillette's case an oligopolist, lacked the incentive to extend the durability of its product until a competitor moved in with a longer-lasting product.

Thus, durability became a contentious issue, and the practices discussed above heightened the interest of academics in durability choice under various market structures: Will durability choice be different under different conditions of competition and monopoly? This was the question Chamberlin first posed in 1953. Empirical or anecdotal evidence seemed to suggest that monopolists would indeed produce goods of shorter durability than competitive markets, and the 1960s witnessed an increase in economic studies that sought to analytically confirm these observations. These are discussed in the following section on the second period in the history of economic thought on durability.

---

[4] A good summary of anecdotal evidence about these practices can be found in Avinger (1981).

3.5 "Flawed Analytic" Period in the History of Economic Thought on Durability

The 1960s witnessed several attempts to expand the traditional price–quantity economic analysis by integrating product durability. The pioneering work of Martin (1962) should be acknowledged as the first to propose an analytic solution to Chamberlin's question of durability choice and market structure.[5] Martin (1962) wondered, "considering the importance of this question, it is surprising that so little has been said about it in the literature."[6] He showed that, under plausible assumptions, a monopolist has an incentive to reduce the durability of the goods its produces below "the length of life that would exist under competitive conditions."

Kleiman and Ophir (1966) followed up on the work by Martin (1962) with a careful discussion of "durability of durable goods." The authors offered a thoughtful discussion of the concept of durability, separating, for example, the choice of durability from the decision to discard or replace an asset, that is, they distinguished between ex ante and ex post durabilities – this was not picked up by later authors.[7] They noted that "a product's length of life differs from other attributes like color, shape, etc. in that preferences for it depend not upon consumers' taste but only upon an economic variable" (Kleiman and Ophir, 1966). Their analysis builds on Martin's (1962); it drops many of his assumptions but still leads to the same conclusion:

> It is obvious that a monopolist will make his profit by restricting the quantity of the service made available to the user [and] will achieve this restriction partly by producing shorter-lived assets than would a competitive industry. (p. 177)

---

[5] It is interesting to note that a solution to Chamberlin's question was attempted almost 10 years after the question was first formulated. There seems to be a time lag or cycles of roughly 10 years in economic analysis of durability, as will be seen later with Swan, Coase, and Bulow.

[6] "What governs durability of different products, and under different conditions of competition and monopoly; and how does the optimum defined by profit maximization compare with standards defined by the public interest or welfare criterion?" (Chamberlin, 1953).

[7] See Section 3.9, point (4), for a discussion of durability in an ex ante and in an ex post sense.

Levhari and Srinivasan (1969) and Schmalensee (1970) offered similar analyses showing that a monopolist would produce goods with lower durability than would exist under competitive conditions. Levhari and Srinivasan (1969) assumed *one-hoss-shay* goods (i.e., having the property of sudden and total failure or breakdown; see Saleh, 2005) and Schmalensee (1970) extended their results to goods that deteriorate geometrically. Although their work built on that of Martin (1962) and Kleiman and Ophir (1966), it is surprising that Levhari and Srinivasan (1969) cite no previous work of relevance to their paper, and Schmalensee (1970) only references Levhari and Srinivasan (1969).

This second period in the history of economic thought on durability attempted to provide an analytic solution to the question posed by Chamberlin on product durability choice under different market conditions. Unfortunately, it is widely recognized, since Swan (1970), that the authors cited in this second period reached their conclusions via flawed analysis. It is ironic, however, that, 10 years later, after Swan's counterintuitive results and assumptions had been carefully examined and to a large extent accepted – just like the situation of the authors in this second period – Swan himself would be proven wrong; our present authors ironically reached the right conclusions via flawed analysis.

3.6 The Swan-Centric Period in the History of Economic Thought on Durability

As discussed in the introduction of this chapter, in 1970, Peter Swan, an Australian economist working at Monash University at the time,[8] published a surprising and counterintuitive result, which occupied many economists for the remainder of the decade: he showed that "the optimal degree of durability for the monopoly is identical with the competitive solution" (Swan, 1970). In other words, manufacturers of durable goods make the same choice of product durability irrespective of whether they are under competitive market conditions or monopoly (Swan 1970, 1971, 1972; Sieper and Swan, 1973).

---

[8] He later held visiting faculty positions at the University of Chicago and Rochester University. As of 2007, he was a chaired professor of economics at the University of New South Wales, Australia.

Swan's theoretical result seemed to challenge not only traditional academic results, but also observed monopolist or oligopolist behavior and the limited empirical evidence on the subject.

Swan showed that a specification error was introduced in the analysis by Levhari and Srinivasan (1969), and consequently by other authors of the 1960s cited in the previous section, namely that the way they calculated the profits of a monopolist were confined to replacement sales and did not include initial sales (Swan, 1970). Had initial sales been included in their analysis, Swan showed that the previous authors would have found, as he did, that "the monopoly and competitive solution to the durability question [are the same]: firms will select the same degree of durability regardless of monopoly power" (Swan, 1972).

Swan's finding is the result of what can be described as steady-state or long-term equilibrium analysis. He posits, for example, just as did all previous authors and many authors after him until perhaps Waldman (1993), an infinite time horizon over which a monopolist sells a quantity Q of the *same durable good* with durability N, *at every N intervals ad infinitum*. This assumption, or more generally this *static or steady-state* analytical framework, is not credible or convincing to anyone who has an understanding of the dynamic nature of technology development and real world durable goods. Swan's result relies on an important (but flawed) assumption of perfect substitution between different vintages of durable goods. In other words, old and new units are perfect substitutes and "additional units of service can restore the quality of the depreciated good, i.e., quantity and quality are perfect substitutes" in Swan's model[9] (Hendel and Lizzeri, 1999).

In addition, Swan found that the cases of a monopolist seller and a monopolist renter are identical: "monopoly profits are identical whether the product is rented or sold" (Swan, 1970). This finding would later be

---

[9] Many authors will later criticize this assumption as unrealistic. Hendel and Lizzeri (1999), for example, argue half-jokingly that "it would be very costly to take a car that has lost half its value and make it as good as new by buying half a new automobile." And Waldman (2003) states, "for most durable goods, [this assumption] is not at all realistic. For products such as automobiles, televisions, refrigerators and toasters, a consumer cannot combine services from a number of used units to create a perfect substitute for a new unit."

exposed as a major flaw in Swan's analysis, as the distinction between a monopolist seller and a renter is significant (further discussed in the following section). Several other assumptions underlie Swan's counterintuitive result, and many economists focused for the remainder of the decade (the 1970s) either on questioning these assumptions or on testing the robustness of Swan's result to relaxing some of these assumptions. "The conclusion of this literature, not surprisingly, is that Swan's findings are robust to relaxing some of his assumptions, but not others" (Waldman, 2003).

It is important to clarify that Swan's results indicate that monopolists, just like firms under competitive market conditions, produce goods at the lowest cost of service and therefore do not distort durability. However, monopolists use their market power to manipulate price rather than durability, or as stated more colorfully by Schmalensee (1979): "a monopoly offers the same product as a competitive industry; its only sin is to charge more."

No other salient results or contributions were made in the 1970s. Two excellent surveys of the economic literature of this period can be found in Schmalensee (1979) and Waldman (2003).

3.7 The Identification of the Time Inconsistency Problem for Durable Goods Monopolists

In the early 1980s, Nancy Stokey (1981) and Jeremy Bulow (1982) brought to light a little-known paper published by Ronald Coase[10] in 1972 that had devastating consequences for Swan's analysis and conclusion. Previous economists seemed either to have been unaware of Coase's work or not to have understood or trusted its implications until Stokey (1981) formally proved Coase's conjecture[11] and Bulow (1982) clarified its implications with

---

[10] A very interesting and unusual economist: he published a limited number of articles, and made very limited use of mathematics. Yet his work had a profound impact on economics. He was born in England in 1910 and moved to the United States in 1951. He taught at the University of Virginia and spent the rest of his career on the faculty at the University of Chicago. He won the Alfred Nobel Memorial Prize in Economic Sciences in 1991 for his work on transaction costs and property rights.

[11] Another modeling and proof of Coase's conjecture in a game theoretic context can be found in Gul et al. (1986).

respect to product durability and monopoly.[12] Coase's paper would occupy economists interested in issues of durability for the rest of the 1980s and form the basis for many other works on the subject in the 1990s.

What was Coase's conjecture? How is it relevant to the problem of durability choice, and what implications does it carry for Swan's analysis and result?

Coase (1972) introduced what later came to be called the *time inconsistency problem* of a durable goods monopolist. In this problem, he integrated two separate considerations with respect to the analysis of durability and monopoly, namely pricing dynamics and consumer rational expectations:

1. **Pricing dynamics**: On one hand, there is the pricing dynamics of a monopolist selling durable goods and attempting to price discriminate over time between consumers with different valuations of the product. Consider the following hypothetical sequence of events: the monopolist first sells the good at the monopoly price to consumers with a high valuation of the product. When the market clears of high-valuation consumers, the monopolist, recognizing residual demand for the product among lower-valuation consumers, will seek to maximize his profits in this second period by producing and selling additional products at a lower price to these consumers (who have a lower valuation of the product). This sequence of events continues until the competitive price, namely the marginal cost, has been reached (Coase, 1972; Bulow, 1982; Kahn, 1986; Fishman and Rob, 2000; Waldman, 2003). "This ability to exploit residual demand exists as long as the stock of the good falls short of the stock that would be produced by a competitive market. The monopolist will accordingly produce until the competitive stock has been achieved" (Bond and Samuleson, 1984).

2. **Consumer rational expectations**: On the other hand, there are rational expectations of consumers who rightly perceive the monopolist's incentive to maximize its profits in every time period by price discriminating

---

[12] Is it fair to make such a statement? Consider the following: to date (June 2006), Coase (1972) has been cited 223 times on the Web of Science Citation Index; 95% of these citations are post-Bulow (1982). Or, stated differently, Coase (1972) was cited on average 1 time per year prior to Bulow (1982), and received on average 9 citations per year in the decades following Bulow (1982). In addition, approximately 70% of the papers that cite Bulow (1982) – 172 as of June 2006 – also cite Coase (1972).

over time between consumers with different valuations of the good. In other words, consumers will expect the hypothetical sequence of events discussed in (1) to unfold. Consequently, even high-valuation consumers will refuse to pay the initial monopoly price and will value the good at the competitive price. "Thus no one will buy at the initial price unless that price is equal to the price that will ultimately prevail, namely marginal cost" (Kahn, 1986).

Coase's major contribution is twofold: first the identification of the time inconsistency problem of a durable goods monopolist (discussed above in (1) and (2)), and second the resolution of – or suggestions for how the monopolist can resolve – the time inconsistency problem, which will be discussed shortly.

But before proceeding further, it is important to highlight the implications of the time inconsistency problem for a durable goods monopolist. The consequence of the time inconsistency problem in the case of a monopolist seller is that "with complete durability,[13] the price becomes independent of the number of suppliers and is thus always equal to the competitive price" (Coase, 1972). In other words, a monopolist seller cannot truly exercise monopoly power. Commenting on this result, Kahn (1986) noted that "the intuition originally proposed by Coase (1972) is that a durable goods monopolist is in fact competitive, because he is in the same market as himself tomorrow."

How can a monopolist avoid the time inconsistency problem (and regain his monopoly power)? Coase proposes three solutions, two of which bear devastating consequences for Swan's analysis and result: by making special contractual arrangements or precommitments (to maintain prices, not to produce or sell additional units, to buy back, etc.), by renting instead of selling, or by reducing the durability of its goods below the socially optimal level, a durable goods monopolist can avoid or mitigate the time inconsistency problem.

1. **Special contractual arrangement or precommitment**: The time inconsistency problem arises because of the consumers' rational expectation

---

[13] Coase uses "complete durability" to mean an infinitely durable good. He uses the example of land, an infinitely durable good, in his 1972 paper to illustrate his discussion.

that the monopolist will keep producing and selling goods at an increasingly lower price, until the competitive price is reached. One way a monopolist can avoid this problem is by credibly committing up front to future prices, or not to produce more units and sell them at a lower price. Coase used the example of a hypothetical monopolist landowner who could commit to hold unsold in perpetuity a quantity of land (in order to avoid the price decline), or who would agree to buy back land at a price just lower than the initial selling price, "thus making it against his interest to sell more than the [initial] land" (Coase, 1972). Some authors have argued that firms cannot credibly commit not to produce or limit their production in the future (Bond and Samuelson, 1984), and therefore another solution to the time inconsistency problem should be sought.

2. **Renting instead of selling**: In his analysis, Swan treated monopolist renters and sellers the same. This proved to be a major flaw. Coase (1972) introduced a necessary difference between monopolist renters and sellers. The difference is related to the consumers' rational expectation of monopolists' incentive to overproduce or not and of how that behavior affects the future value of the goods (rented or purchased). Ultimately, the difference can be reduced to the question of whether the monopolist internalizes the losses from its future behavior (e.g., overproducing), or whether these losses are borne by the consumer. In his landowner monopolist example, Coase clarifies that "by selling more land, [the landowner monopolist] would reduce the value of land now owned by those who had previously bought from him – but the loss would fall on them, not him." Bulow (1982) put it more forcefully by stating, "if a seller over-produces, the losses are suffered by old purchasers in whose welfare the seller has no direct interest. The loss is not internalized" (Bulow, 1982). Therefore, in the monopolist seller's case, consumers with rational expectations will perceive the monopolist's incentive to further produce and sell at increasingly lower prices until the competitive price is reached, and will therefore refuse to pay the initial high price (Coase, 1972; Bulow, 1982, 1986; Waldman, 2003). This is an implication of the time inconsistency problem, which arises because the monopolist seller does not internalize the adverse effects from its future

behavior.[14] The case of a monopolist renter is very different. Coase intuited that a monopolist renter can overcome the time inconsistency problem by "leasing for relatively short period of time" instead of selling. It would therefore not be in "his self interest [...] to lease more by charging lower prices" (Coase, 1972), because he retains ownership of the products and the potential losses would be *internalized* in the case of a monopolist renter (Waldman, 1993). In other words, a monopolist can avoid the time inconsistency problem by renting instead of selling. Coase's treatment of this subject was very informal, and it was not until Stokey (1981) published her work that Coase's original intuition was proven. In addition, Bulow (1982) proved that a monopolist seller "will end up over-producing relative to the renter."

But there are cases in which renting is not possible; for example, monopolists may be required by law to sell rather than rent their products. "Renting may be ruled out for legal reasons. IBM, Xerox, United Shoe Company, all began by only renting their products but are now required to also make sales" (Bulow, 1982). Coase proposed a third solution that would allow monopolist sellers to mitigate the time inconsistency problem: reduced durability!

3. **Reduced durability**: Coase recognized a third alternative that mitigates the time inconsistency problem and restores its ability to charge higher prices, even though that alternative was not available to his monopolist landowner: to make the good less durable.[15]

   The production of a less durable good as against a more durable good is very similar to the policy of leasing since, by making the good less durable, the producer sells the services provided by the good for short periods of time (because the good wears out) whereas in leasing the same result is achieved by selling the services provided by the good in short period segments. [...]. Another circumstance reinforces the conclusion that making a good less

---

[14] Said differently, "once the monopolist sells a machine, he is no longer interested in what happens to the value of the machine. However, his customers presumably realize this and take this factor into account in determining how much they are willing to pay for a machine" (Bulow, 1982).

[15] The durability of the good in the case of the landowner, land, is infinite (or "complete") and is not dependent on the owner's choice.

durable will enable the monopolist producer to charge a higher price [and thus mitigate the time inconsistency problem]: what a consumer has to fear is an increase in supply during the period in which he is deriving services from the good [hence the consumer's unwillingness to pay the initial higher price]. The less durable the good, the shorter is this period. Lessened durability reduces the gain from an increase in supply and thus reduces the likelihood that it will occur. (Coase, 1972, pp. 148–49)

Therefore, monopolist will tend to produce lower durability and charge higher prices. In contrast to Swan's result that monopolist distorts price only, not durability, Coase argued that because of the time inconsistency problem, which Swan had not considered, a monopolist will distort both price and durability (compared with a competitive industry) in order to restore monopoly power. Again, however, as in the case of renting versus selling, Coase's treatment of reduced durability as a strategy to mitigate the time inconsistency problem was very informal, and it was not until Bulow (1982, 1986) that the implication of Coase's conjecture regarding reduced durability, or as Bulow calls it *planned obsolescence*, became widely accepted. Keep in mind, though, that reduced durability is just one way a monopolist can mitigate the time inconsistency problem.

In summary, this fourth period in the history of economic thought on durability saw the recognition of the time inconsistency problem of a durable goods monopolist. This problem was first identified by Coase (1972) and brought to light and formally proved by Stokey (1981) and Bulow (1982). The implications of the time inconsistency problem, undermined Swan's analysis and result: that monopolist renters and sellers are identical, and that the durability choice is independent of market structure (i.e., producers make the same choice of product durability irrespective of whether they are under competitive market conditions or monopoly). Swan's solution to the durability choice "was not dynamically consistent" (Bulow, 1986). In other words, it did not take into account the time inconsistency problem and therefore was incomplete and incorrect.

Coase's paper would occupy economists interested in issues of durability for the remainder of the decade (the 1980s) and form the basis for many other works on the subject well into the 1990s. It gives one pause for thought that 10 years after Swan proved the 1960s authors wrong (Martin, 1962; Kleiman and Ophir, 1966; Levhari and Srinivasan 1969; Schmalensee, 1970), he would himself be proven wrong, and it would be recognized that these early authors ironically had reached the right conclusions via flawed analysis.

3.8 Recent Economic Literature on Durability

Recent literature on durability has developed in a number of directions, either by relaxing some of the assumptions underlying the Coase–Stokey–Bulow analyses, or by expanding the analysis framework and making it more relevant to real world goods and markets. The following paragraphs provide a brief description of a few papers from this literature, particularly in connection with the role of replacement sales and secondary markets and the introduction of new goods.

Bond and Samuelson (1984) studied the impact of product depreciation and replacement sales on a monopolist's choice of product durability. The authors also explored a form of precommitment (discussed previously in (A)) that would allow the monopolist to overcome the time inconsistency problem, which is for the firm to credibly constrain its production capacity, "perhaps by installing a plant size which is limited and adjusted only at a prohibitive expense" (Bond and Samuelson, 1984). Their results are consistent with the Coase–Stokey–Bulow analyses and findings.

Kahn (1986) took issue with Coase's claim that, in the absence of precommitment, a monopolist seller would act competitively, not in the long run but "in the twinkling of an eye" (Coase, 1972). Kahn first discussed the need for a continuous time framework to test whether Coase's conjecture holds and under what assumptions.[16] He then showed by relaxing two assumptions underlying Coase's result, namely the no barrier to transaction or the length of the trading period being zero[17] and constant marginal cost of production assumptions, that the time inconsistency problem facing a monopolist seller is not as acute as Coase suggested and does not lead

---

[16] As opposed to a discrete analysis of the problem, for example, Bulow's (1982) two-period model.

[17] "The consumer will regard the goods sold by the monopolist tomorrow as perfect substitute for those sold today only if there is no loss in waiting until tomorrow" (Kahn, 1986).

to efficient or competitive behavior. "The durable goods monopolist produces less than the efficient producer in every period; but more than the pre-committed monopolist" (Kahn, 1986). He does not, however, address the choice of reduced durability as a means of mitigating the time inconsistency problem (irrespective of how severe this problem is).

Bulow (1986) proposed an economic theory of planned obsolescence, which he defined as the "production of goods with uneconomically short lives so that customers will have to make repeat purchases." In this, he follows the tradition of using durability as a proxy for obsolescence, but recognizes, however, that this is "the greatest weakness of [his] paper." Bulow finds that monopolists will produce goods with "inefficiently short lives" and that oligopolists have an incentive to collude to reduce durability below competitive level. He also shows that, as a monopolist's markets become more competitive, the firm has an incentive to steer customers toward purchase rather than rent and increase its sales-to-rent ratio. One important novelty in Bulow (1986) is his finding that, under certain technology and market conditions, oligopolists or monopolists facing threats of future entry may choose increased durability (above competitive level) as a strategy to deter entry of potential competitors.

Waldman (1993) discussed planned obsolescence in the context of new product introduction, as opposed to reduced durability. He showed, using a model that assumes infinitely durable goods, that a monopolist has a high "incentive to introduce new products that make old units obsolete." Waldman also distinguishes between monopolist renters and sellers and uses an argument similar to the Coase–Stokey–Bulow analyses to show that the reason a monopolist seller has incentive to introduce a new products is that "the firm [does] not internalize how its current behavior affects the value of units previously sold." Similarly, Choi (1994) discussed the case of a monopolist who introduces a new generation of products incompatible with the old ones. In some cases, the author shows that this will accelerate obsolescence and dissatisfaction of consumers with the old products "and will enable the monopolist to extract more surplus from old consumers." Building on these two works – Waldman (1993) and Choi (1994) – Waldman (1996) analyzed the problem of a monopolist's decision to invest in R&D, "in this way improving the quality of what it will sell in the future, [which] has the potential to reduce the future value of current and past output." He finds, using a simplistic model for R&D impact on product improvement, that in the absence of an ability to commit to future value of R&D, a monopolist seller faces a time inconsistency problem similar to the one identified by Coase (1972). "The result is that the [monopolist seller] chooses an investment in R&D greater than the amount that maximizes its own profitability [in the long run]" (Waldman, 1996), which is beneficial from the standpoint of social welfare.[18]

Most of the literature has associated a negative connotation with a shorter durability (below the efficient level) and assumed it to be *wasteful* in a social sense. Fishman et al. (1993) proposed a different perspective on the subject. They argued that excessive durability can delay the development of new technologies and serve as a barrier against the development of new products (somewhat similar to the argument of Bulow (1986) of long durability as a barrier to deter entry of potential competitors). The authors argued that planned obsolescence, that is, shorter durability below the efficient level, can promote technological advancement and "may be preferred to long lasting products and slow innovation," which in turn can lead to a stagnating economy.

Hendel and Lizzeri (1999) investigated both the incentives of monopolists to interfere with secondary markets for used goods and the various ways that such firms can do so. A monopolist may seek to control secondary markets and use them "as a tool for extracting more surplus from consumers." Hendel and Lizzeri show, for example, that a monopolist has an incentive to prevent or restrict consumers' ability to maintain their durable goods.[19] In addition, the authors take issue (rightfully so) with Swan's perfect

---

[18] Although Waldman (1996) makes an important and thought-provoking contribution, his paper is ambiguous at times, and could have benefited from a clearer discussion of the issues at stake. Consider, for example, the following: "As a result, the private incentive to invest in R&D is less than the incentive that is social welfare maximizing. What happens, therefore, is that in the absence of the ability to commit to a future value of R&D, the time inconsistency problem that increases the monopolist's incentive to invest serves to offset the incentive for investment that is too low in the case where the firm has the ability to commit" (Waldman, 1996).

[19] "If consumers are allowed to maintain the goods as they wish, the appeal of used goods for the marginal consumer of the new good would increase. This lowers this consumer's willingness to pay for the new good" (Hendel and Lizzeri, 1999).

substitution assumption, that is, that "different vintages of the goods are perfect substitutes for each other." This assumption, they show, is not appropriate for many goods and does not allow for an active role and analysis of secondary markets. The authors' premise is that a monopolist faces consumers with different valuations for quality, and different vintages of a durable good appeal to different consumers; that is, "consumers have heterogeneous valuation for quality." In this context, the authors introduce what they termed the "substitution effect" in the monopolist's choice of durability; that is, new and used goods are imperfect substitutes: "the durability of the product [indicates] how close a substitute the used good is going to be for future units of the new good." They find that "in contrast to Swan's famous independence result, a monopolist does not provide socially optimal durability; allowing the monopolist to rent does not restore socially optimal durability [either]."

Several authors have addressed issues of durable goods pricing under monopoly conditions. Major contributions were made by Mussa and Rosen (1978), and more recently by Fudenberg and Tirole (1998), to name a few. These works, however, were not concerned with durability choice and therefore will not be discussed here. The interested reader is referred to Waldman (2003, 2004) for a thorough literature review on durable goods that is not restricted to the durability choice problem.[20]

3.9 Limitations of Current Economic Thinking about Durability

Despite the significant progress made in our understanding of durability choice and market structure since Swan and Coase, major limitations still remain unaddressed in the study of durability. The following paragraphs describe some of these limitations, and, unlike Schmalensee's warning (from the opening paragraph of this chapter), they constitute an invitation

---

[20] Waldman argues that Akerlof's (1970) work on adverse selection under asymmetric information between buyers and sellers is a major contribution to durable goods theory as well. I was not able to find papers that integrate durability choice and asymmetric information. However, two recent papers, Hendel and Lizzeri (2002) and Johnson and Waldman (2003), address the role of leasing and buy-backs of durable goods under asymmetric information.

for further research into this area still rich for theoretical and empirical contributions.

1. The economic literature has mainly focused on the durability of consumer goods. Little attention, if any, has been given to intermediate (Waldman, 2003) or capital goods. The issues pertaining to durability of capital goods, complex engineering systems, or even infrastructure are more complex and multidisciplinary in nature than the study of durability of consumer goods. The current economic thinking about durability choice is focused on profit maximization from sales of durable goods and does not include, for example, the potential for revenue generation from servicing the asset in the case of a capital good. Current economic analysis of durability choice may therefore be of limited applicability to the case of intermediate or capital goods.

2. The economic literature has addressed the choice of product durability under monopoly or (perfect) competition conditions. There are no studies, however, that look at the durability choice from the customer's perspective: if the customer had market power – to make it formal, call this the durability choice under monopsony – how would he or she choose the durability of the asset to be acquired? And how would this "optimal" durability choice from a customer's perspective compare with the "socially optimal durability"?

3. This point is closely related to the previous one. The current economic literature is replete with discussions of "optimal durability." But the concept of optimality is of course meaningless without an a priori notion of a metric with respect to which optimality is sought. Metrics are essential for decision-making and help guide design choices (e.g., product durability). But what is optimal according to one metric is not likely to be optimal given another metric. Unfortunately, there is limited discussion of metrics, or optimality with respect to what, in the economic literature on product durability. Some authors, however, have been more precise in their use of this attribute (optimal) and write about "socially optimal durability"; that is, optimality is considered with respect to the durability choice made by a producer under competitive market conditions. This is a fair consideration and significant progress has

been made in this area, because it has been the main focus of the economic literature on durability to date. But one can posit other metrics according to which product durability can be sought to be optimal. For example, although "socially optimal durability" implies a population of durable goods, one can adopt instead a monadic perspective on durability and raises a new set of optimal durability questions: consider, for instance, one durable good (or capital good) and think of it as a value delivery artifact providing a flow of service over time. In addition, assume a nonzero marginal cost of durability. Then one can ask: how should durability of this one good be selected, and what needs to be taken into account in order to maximize its expected net present value (think of a complex engineering system, instead of a light bulb)? And how does this "optimal durability" that maximizes one system's net present value compare with the "socially optimal durability"? The difference in perspective here provided on the choice of durability is one between a *monadic* approach and the traditional *population*[21]-*based* approach.

4. Another major limitation in the economic literature on durability is the use and misuse of the word, and the concept, of durability itself. There are in fact three related but distinct concepts underlying the notion of durability or an artifact's relationship with time, but they are unfortunately seldom distinguished in the literature. The first one – call it *design lifetime – is durability in an* ex ante *sense,* that is, the requirement that the manufacturer selects for the *intended duration of operation of the system* – alternatively, this requirement can be imposed on the producer by the market, the customers, or the regulators.[22] Designers and engineers will make different design choices for the system's structure and components depending on whether the selected design lifetime is, for example, 1 year, 5 years, or 10 years. A related but different concept that captures another aspect of an artifact's relationship with time is

---

[21] Of durable goods.

[22] It should also be noted that *durability* and *design lifetime* sometimes carry different connotations that deserve to be recognized: durability has often been used to characterize the durability of a consumer good (product of limited complexity), whereas design lifetime is often associated with complex engineering systems or capital goods.

durability in an ex post *sense*. It is an *observation*, in the sense of a direct measurement of how long the product actually remains operational before it breaks down. It is related to the design lifetime, the intended duration of operation of the product, but is conceptually different, and of course need not be the same, just as any estimation can be different from an observation (in a statistical sense). In addition, durability in an ex post sense in associated with the notion of product failure and has physical integrity connotations (i.e., the artifact is "broken").[23] A third concept is required to clearly discuss issues of durability and planned obsolescence: call it *service life* or *economic life*, that is, the actual duration of operation before the product or system is retired. A durable (capital) good may be retired for economic reasons – its services are not longer required, or better competing services are available more cost-effectively – as it becomes functionally obsolete, although it can still be operational (i.e., it is not "broken" but no longer needed). Notice the difference between durability in an ex post sense and service life. They are conceptually different and are driven or determined by different considerations: the former is associated with physical failure (engineering choices and product usage), whereas the latter is related to functional inadequacy[24] (evolving market needs, competition, and inability to adapt to change). These three concepts, design lifetime, durability in an ex post sense, and service life, are traditionally lumped into one word, durability, in the economic literature. These three concepts allow a clearer discussion of "durability" issues and raise a whole new set of interesting research questions with economic, managerial, and engineering underpinnings and implications.

5. Another important limitation in the economic literature on durability is the treatment of durability, in any of the above senses, as a deterministic

---

[23] See the discussion in the Epilogue.

[24] The end of each period, durability in an ex post sense and service life, is signaled by an event, physical failure or time to failure in one case, and obsolescence and time to retirement in the other case. The two events can, but need not, be related. In addition, the occurrence of one event means that the other need not occur (e.g., if a system is retired, it need not be broken nor will it have any longer a chance to break while in operation). This may be one difficulty in identifying these two separate concepts, durability in an ex post sense and service life.

concept, when it is in fact fundamentally a probabilistic or stochastic variable. Valuable insights into the choice of durability (ex ante) may be obtained by adopting a probabilistic perspective – and analytical methods – in investigating issues of durability. Furthermore, it is always assumed that *one choice of durability* has to be made *before the product is designed* and fielded. It is possible, however, to conceive of a design and a scenario by which the potential for multiple durability choices can be embedded in the design of the product. The durability choice can thus be staged over time, and the choice of a specific durability can be postponed until after the product is fielded and some uncertainty resolved. In other words, there would be different points in time at which different durability choices can be made (for example, to extend the design lifetime of the system as it is, to upgrade its capabilities, or to modify them). This point is related to the first one discussed in this section on the possibility of maintaining and servicing a capital good. The financial literature on real options has explored similar concepts in project developments; see for example Trigeorgis (1996) or Amram and Kulatilaka (1999). It would be interesting to investigate the implication of a similar mindset in the durability choice of durable or capital goods.

In summary, despite the significant progress made in durable goods theory and in our understanding of durability choices under different market structures, many limitations still remain unaddressed in the current economic thinking about durability. It is hoped that the limitations discussed above be viewed as an invitation to further research into this area, as significant opportunities still remain to make important theoretical and empirical contributions to this subject.

3.10 Conclusions

This chapter provided a narrative of the development of economic thought on durability. Four periods were proposed to communicate the dynamics and evolution of the history of the durability choice problem in the economic literature: (1) the origins and preanalytic period with Knut Wicksell and Edward Chamberlin as the main protagonists, (2) the "flawed analytic" period, (3) the Swan-centric period, and (4) the identification of the time inconsistency problem for durable goods monopolists, or the Coase (1972) and post-Coase period. Historians often engage in such periodization exercises, i.e., the division of history into periods. The exercise can be useful in helping conceptualize and communicate certain dynamics or thematic threads, but *periods* should not be taken too rigidly or literally, as they are rarely linear and often overlap. Nevertheless, the narrative here provided showed a clear pattern of rise then stumbling in economic thought on durability, followed by a more subtle and careful development of an economic theory of durability choice.

Following an exposition of the historical development of economic thought on durability, this chapter discussed the many limitations that still remain unaddressed in this literature. It is hoped that these limitations will be viewed as an invitation for researchers to further explore this area still rich for theoretical and empirical contributions.

The following analogy with biology may be interesting or amusing: evolutionary biology is sometimes described as "evolution by jerks," that is, long periods of equilibrium punctuated by moments of acute change. The economic theory of durability seems to have exhibited a similar pattern of evolution, with time lags or cycles of roughly 10 years: from Chamberlin (1953) to Martin (1962) to Swan in the early 1970s, Stokey and Bulow in the early 1980s, and Waldman (1993).

APPENDIX – ORIGINS OF COASE'S CONTRIBUTION TO THE TIME INCONSISTENCY PROBLEM OF DURABLE GOODS MONOPOLISTS

Coase's (1972) contribution lies at the intersection of two histories of economic thought, the first dealing with the inconsistency of optimal plans in general, and the second related to durable goods theory (discussed in this chapter).

Although Coase does not cite any references in his 1972 paper, his work builds on the general time inconsistency problem first identified by Robert Strotz in 1955 in a paper titled "Myopia and Inconsistency in Dynamic Utility

Maximization," and later further investigated by Robert Pollak in 1968 in "Consistent Planning." This same economic thread led to the very influential article by Kydland and Prescott (1977), "Rules Rather Than Discretion: The Inconsistency of Optimal Plans," in which the authors show that optimal control theory cannot be made applicable to economic planning when there are economic agents with rational expectations. Coase (1972) appears to be an application of Strotz's insights to the durable goods monopolist problem of durability choice. Strotz (1955) formulated the time inconsistency problem as follows, and suggested a strategy of precommitment as a way of avoiding this problem:

> An individual is imagined to choose a plan of consumption for a future period of time so as to maximize the utility of the plan as evaluated at the present moment.... Our problem arises when we ask: if he is free to reconsider his plan at later dates, will he abide by [his original plan] or disobey it?... Our answer is that the optimal plan of the present moment is generally one which will not be obeyed, or that the individual's future behavior will be inconsistent with his optimal plan.... If the inconsistency is recognized, the rational individual ... may pre-commit his future behavior by precluding future options so that it will conform to his present desire as to what it should be. (p. 165)

REFERENCES

Akerlof, G. "The market for 'lemons': Quality uncertainty and the market mechanism." *Quarterly Journal of Economics*, 1972, 83 (3), pp. 488–500.

Amram, M., and Kulatilaka, N. *Real Options: Managing Strategic Investments in an Uncertain World*. Harvard Business School Press, Boston, 1999.

Avinger, R. L., Jr. "The Economics of Durability." Ph.D. dissertation. Duke University, 1968.

Avinger, R. L., Jr. "Product durability and market structure: Some evidence." *Journal of Industrial Economics*, 1981, 29 (4), pp. 357–74.

Ben-Akiva, M., and Gopinath, D. "Modeling infrastructure performance and user costs." *Journal of Infrastructure Systems*, 1995, 1 (1), pp. 33–43.

Bond, E., and Samuelson, L. "Durable good monopolies with rational expectations and replacement sales." *RAND Journal of Economics*, 1984, 15 (3), pp. 336–45.

Bright, A. A. *The Electric Lamp Industry*. Macmillan, New York, 1949.

Bulow, J. "Durable-goods monopolists." *Journal of Political Economy*, 1982, 90 (2), pp. 314–332.

Bulow, J. "An economic theory of planned obsolescence." *Quarterly Journal of Economics*, 1986, 101 (4), pp. 729–49.

Chamberlain, E. H. "The product as an economic variable." *Quarterly Journal of Economics*, 1953, 67 (1), pp. 1–29.

Choi, J. P. "Network externalities, compatibility choice, and planned obsolescence." *Journal of Industrial Economics*, 1994, 42 (2), pp. 167–82.

Coase, R. H. "Durability and monopoly." *Journal of Law and Economics*, 1972, 15 (1), pp. 143–9.

Fishman, A., Gandal, N., and Shy, O. "Planned obsolescence as an engine of technological progress." *Journal of Industrial Economics*, 1993, 41 (4), pp. 361–70.

Fishman, A., and Rob, R. "Product innovation by a durable-good monopoly." *RAND Journal of Economics*, 2000, 31 (2), pp. 237–52.

Fudenberg, D., and Tirole, J. "Upgrades, trade-ins, and buy-backs." *RAND Journal of Economics*, 1998, 29 (2), pp. 235–58.

Gul, F., Sonnenschein, H., and Wilson, R. "Foundations of dynamic monopoly and the Coase conjecture." *Journal of Economic Theory*, 1986, 39 (1), pp. 155–90.

Hendel, I., and Lizzeri, A. "Interfering with secondary markets." *RAND Journal of Economics*, 1999, 30 (1), pp. 1–21.

Hendel, I., and Lizzeri, A. "The role of leasing under adverse selection." *Journal of Political Economy*, 2002, 110 (1), pp. 113–43.

Johnson, J., and Waldman, M. "Leasing, lemons, and buy-backs." *RAND Journal of Economics*, 2003, 34 (2), pp. 47–65.

Kahn, C. "The durable goods monopolist and consistency with increasing costs." *Econometrica*, 1986, 54 (2), pp. 275–94.

Kleiman, E., and Ophir, T. "The durability of durable goods." *Review of Economic Studies*, 1966, 33 (94), pp. 165–78.

Kydland, F. E., and Prescott, E. C. "Rules rather than discretion: The inconsistency of optimal plans." *Journal of Political Economy*, 1977, 85 (3), pp. 473–91.

Levhari, D., and Srinivasan, T. N. "Durability of consumption goods: Competition versus monopoly." *American Economic Review*, 1969, 59 (1), pp. 102–7.

Martin, D. "Monopoly power and the durability of durable goods." *Southern Economic Journal*, 1962, 28 (3), pp. 271–7.

Mussa, M., and Rosen, S. "Monopoly and product quality." *Journal of Economic Theory*, 1978, 18 (2), pp. 301–17.

Pollak, R. A. "Consistent planning." *Review of Economic Studies*, 1968, 35 (2), pp. 201–8.

Saleh, J. H. "Perspectives in design: The deacon's masterpiece and the hundred-year aircraft, spacecraft, and other complex engineering systems." *Journal of Mechanical Design*, 2005, 127 (5), pp. 845–50.

Schamlensee, R. "Regulation and the durability of goods." *Bell Journal of Economics and Management Science*, 1970, 1 (1), pp. 54–64.

Schmalensees, R. "Market structure, durability, and quality: A selective survey." *Economic Inquiry*, 1979, 17 (2), pp. 177–96.

Sieper, E., and Swan, P. L. "Monopoly and competition in the market for durable goods." *Review of Economic Studies*, 1973, 40 (3), pp. 333–51.

Stocking, G. W., and Watkins, M. W. *Cartels in Action*. The Twentieth Century Fund Press, New York, 1946.

Stokey, N. L. "Rational expectations and durable goods pricing." *Bell Journal of Economics*, 1981, 12 (1), pp. 112–28.

Stokey, N. L. "Learning by doing and the introduction of new goods." *Journal of Political Economy*, 1988, 94 (4), pp. 701–17.

Strotz, R. H. "Myopia and inconsistency in dynamic utility maximization." *Review of Economic Studies*, 1955, 23 (3), pp. 165–80.

Swan, P. L. "Durability of consumption goods." *American Economic Review*, 1970, 60 (5), pp. 884–94.

Swan, P. "The durability of goods and regulation of monopoly." *Bell Journal of Economics and Management Science*, 1971, 2 (1), pp. 347–57.

Swan, P. L. "Optimum durability, second-hand markets, and planned obsolescence." *Journal of Political Economy*, 1972, 80 (3), pp. 575–85.

Trigeorgis, L. *Real Options: Managerial Flexibility and Strategy in Resource Allocation*. MIT Press, Cambridge, MA, 1996.

Waldman, M. "A new perspective on planned obsolescence." *Quarterly Journal of Economics*, 1993, 108 (1), pp. 273–83.

Waldman, M. "Durable goods pricing when quality matters." *Journal of Business*, 1996, 69 (4), pp. 489–510.

Waldman, M. "Durable goods theory for real world markets." *Journal of Economic Perspectives*, 2003, 17 (1), pp. 131–54.

Waldman, M. "Anti-trust Perspectives for Durable-Goods Markets." CESifo working paper 1306. CESifo Venice Summer Institute, 2004.

Wicksell, K. *Lectures on Political Economy*. Routledge and Kegan Paul, London, 1934.

4 Analysis of Marginal Cost of Durability and System Cost per Day

PREVIEW AND GUIDE TO THE CHAPTER

Analysis of the marginal cost of durability is a prerequisite for addressing the durability choice problem. This straightforward observation is often forgotten in the economic literature on durable goods, or dismissed by making the unrealistic assumption that the marginal cost of durability is zero. The details of a marginal cost of durability analysis are system-specific. A satellite example is discussed to illustrate the essence and provide a quantitative application of such analysis. First, the impacts of the durability requirement on different subsystems on board the spacecraft are quantitatively explored, that is, how the different subsystems (power, propulsion, thermal, etc.) scale with increased durability; second, the effects of durability on the different subsystems are integrated and typical satellite cost profiles are provided as a function of durability. The result represents the marginal cost of durability of a satellite. In addition to the satellite example, two related metrics are introduced, the cost elasticity of durability and the cost per day, which allow a clear understanding and visualization of the economies of scale, if any, that result from extended durability. This chapter contributes a necessary first step toward a rational choice of durability for engineering systems from a customer's perspective.

4.1 Introduction

The first chapter introduced two broad categories of questions related to an artifact's durability: the technicalities of durability, and the choice of durability. There is in fact a third category of questions, which builds on the first set of questions and is a prerequisite to the second set, the durability choice questions. This overlooked set of questions will be referred to as the marginal cost of durability. Increased product durability, like increased reliability, often requires additional resources (e.g., use of more expensive components or thicker layers of materials) and therefore comes at a cost. The relationship between incremental product durability and the incremental cost required to achieve it, if any, is what is referred to as analysis of marginal cost of durability and is the focus of the present chapter.

This chapter is organized as follows: Section 4.2 provides a brief discussion and definition of durability and some closely related concepts. Section 4.3 makes the case that, because infinitely durable goods do not exist, choices and tradeoffs are necessary to identify the optimal durability for an engineering artifact. A brief discussion of values and metrics is provided before the case is made that the analysis of the marginal cost of durability is a prerequisite for capturing the expected net present value of a system and is therefore a necessary first step in identifying and selecting the optimal durability of an engineering artifact, from a customer's perspective. The details of the marginal cost of durability are system-specific. To illustrate the application of such an analysis, a satellite example is provided in Section 4.4, and the impacts of the durability requirement on the different subsystems on board the spacecraft are quantitatively explored (that is, how the different subsystems, e.g., power, propulsion, thermal, scale with increased durability). The various effects of durability on the different subsystems are then integrated, and typical satellite cost profiles are provided as a function of durability. The example is meant to illustrate the essence and provide a quantitative application of marginal cost of durability analysis. Section 4.5 introduces the cost elasticity of durability metric, and Section 4.6 discusses the cost per day metric, which allows a clear understanding and visualization of the economies of scale, if any, that result from extended durability. Section 4.7 concludes this chapter.

4.2 Nomenclature: Durability, Design Lifetime, and Service Life

To discuss any subject matter clearly, it is important to begin with a clear set of definitions. Much in fact can be gained through careful definitions of terms alone. Durability has several synonyms that are often used interchangeably: for example, a product's or a system's lifespan, design lifetime, service life, or mission duration. There are, however, important but subtle differences among these terms.

There are in fact three related but distinct concepts underlying the notion of durability or an artifact's relationship with time, but they are unfortunately seldom distinguished in the literature. The first one – call it *design lifetime* – is durability in an ex ante *sense*, that is, the requirement that the manufacturer selects for the *intended duration of operation of the system* – alternatively, this requirement can be imposed on the manufacturer by the market, the customers, or the regulators. Designers and engineers will make different design choices for the system's structure and components depending on whether the selected design lifetime is, for example, 1 year, 5 years, or 10 years.

A related but different concept that captures another aspect of an artifact's relationship with time is *durability in an* ex post *sense*. It is an *observation*, in the sense of a direct measurement of how long the product actually remains operational before it breaks down. It is related to the design lifetime, the intended duration of operation of the product, but is conceptually different, and of course need not be the same, just as any estimate can be different from an observation (in a statistical sense). In addition, durability in an ex post sense in associated with the notion of product failure and has a physical integrity connotation (i.e., the product is "broken").[1]

A third concept is required to clearly discuss issues of durability and their corollary, planned obsolescence: call it *service life* or *economic life*, that is, the actual duration of operation before a product or system is retired. A durable (capital) good may be retired for economic reasons – its services are no longer required, or better competing services are available more cost-effectively – as it becomes functionally obsolete, although it can still be operational (i.e., it is not "broken," but it is no longer needed). Notice the difference between durability in an ex post sense and service life. They are conceptually different and are driven or determined by different considerations:

---

[1] Recall, for example, the deacon's one-hoss shay, which breaks down completely after exactly 100 years (Saleh, 2005). This is the physical or structural integrity connotation that is associated with durability in an ex post sense.

the former is associated with physical failure (engineering choices and product usage) whereas the latter is related to functional inadequacy[2] (evolving market needs, competition, and inability to adapt to change).

This chapter deals with durability in an ex ante sense, that is, the intended or design requirement for the duration of operation of a product or a system. The analysis here investigated, namely the marginal cost of durability, should be interpreted as a truncation of the longer marginal cost of ex ante durability or design lifetime.

4.3 On Values, Metrics, and Tradeoffs in the Search for Optimal Durability

Chamberlin (1953) had already remarked that "probably most products not consumed in a single use could be made more durable than they are at somewhat higher cost." This observation captures an important feature of the marginal cost of durability analysis: how does a product's cost scale as a function of its durability requirement? As mentioned earlier, the relationship between the incremental product durability and the incremental cost required to achieve it, if any, is the essence of the analysis of marginal cost of durability.

It is worth emphasizing that the analysis of the marginal cost of durability is a combination of both engineering and cost estimate analyses.

In the following, the case is made that, because infinitely durable products do not exist, choices and tradeoffs need to be made in order to select their durability. After a brief discussion of the need for metrics in decision-making in general, and engineering design in particular, the concept of the expected net present value of a system as a function of its design lifetime is introduced, and it is shown why the analysis of the marginal cost of durability is a prerequisite for identifying the optimal choice of durability of a system.

---

[2] The end of each period, durability in an ex post sense and service life, is signaled by an event, physical failure or time to failure in one case, and obsolescence time to retirement in the other case. The two events can, but need not, be related. In addition, the occurrence of one event means that the other need not occur (e.g., if a system is retired, it need not be broken nor will it any longer have a chance to break while in operation). This may be one difficulty in distinguishing these two separate concepts, durability in an ex post sense and service life.

4.3.1 Choices and Tradeoffs in Lieu of "Infinite Durability"

Even though many economists use the concept in their analyses and models, infinitely durable components or systems do not exist. Failure will occur. It can, however, be postponed or delayed in a number of ways. For example, more durable and reliable components (but also more expensive) can be used instead of generic off-the-shelf components, thus in effect delaying the point of failure. Highway durability, for example, can be increased in several ways, "including better materials, drainage, and construction techniques" (Small and Winston, 1988). But the most common way of increasing highway durability is using thicker pavements or base materials to withstand accumulated damage over a longer period of time. Increased pavement thickness provides additional durability to the highway, but also implies the use of more material and is therefore more expensive than a shorter-lived highway.

More generally, consider a complex engineering system, and think of it as a value-delivery artifact providing a flow of service over time. Increased system durability or design lifetime can be mapped into an additional flow of service or utility that can be derived from the system. Consequently, increased durability can result in an incremental value of the system.

What emerges from the previous observations is that because infinitely durable components or systems do not exist, durability specification requires choices and tradeoffs. System designers or customers, in deciding how much durability is needed, must assess how much durability is worth (the expected present value of durability) and how much they are willing to "pay" for it (the cost of durability). This is illustrated in Figures 4.1a and 4.1b and further discussed in the following section.

4.3.2 On Metrics in Decision-Making and Engineering Design

How one makes choices and tradeoffs is dependent on the notion of metric. Metrics are essential for decision-making. They allow us to characterize and rank different options and provide some guidance in most of our actions and activities. But what is a metric? A metric can be loosely defined as a standard of measurement. It can be measured directly or estimated indirectly, qualitatively or quantitatively, or it can be calculated deterministically

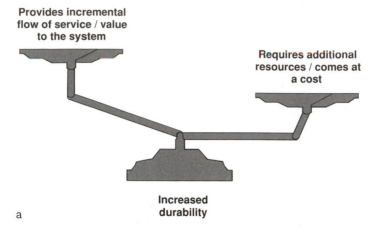

Provides incremental flow of service / value to the system

Requires additional resources / comes at a cost

a

Increased durability

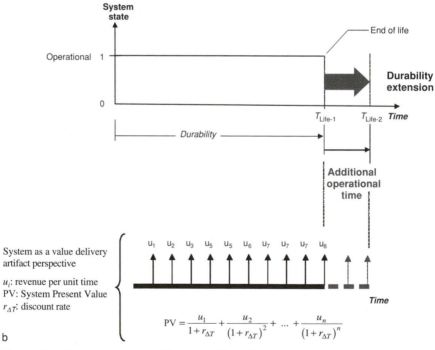

Figure 4.1. (a) Increased durability tradeoff: additional durability comes at a cost, but it enables additional utility to be derived from the system. How much durability is needed requires an assessment of how much it is worth and how much it costs. (b) Extended durability results in additional operational time, which can be mapped into incremental value to the system.

or probabilistically by combining different measurements. In engineering design, metrics play a critical role in guiding design choices. System design and optimization, for example, hinge on the notion of a metrics, or, as is more familiar to the optimization community, on objective or cost functions. Furthermore, what is optimal according to one metric is not likely to be optimal given another metric. For example, an airplane designed and optimized for maneuverability will be very different from one optimized for endurance; conversely, the former will be suboptimal if used for endurance.

How is this relevant to our discussion of the durability choice? To identify and select the "optimal" durability of a system, or to clearly discuss issues of durability choice, it is important to first specify what metric is to be optimized through this particular choice of durability. Unfortunately, although the economic literature is replete with discussions and analyses of "optimal durability," limited attention is given to the metric that is being optimized through this choice of durability. Upon further investigation, it appears that economists have been interested mainly in the choice of durability that maximizes the profits of the manufacturer (under monopolistic or competitive market conditions). This is a fair consideration, and significant progress has been made in this area, because it has been the main focus of the economic literature on durability to date. But one can posit other metrics according to which product durability can be sought to be optimal. For example, instead of the prevalent manufacturer-centric perspective on durability among economists, one can look at the durability choice problem from the customer's perspective (to make it formal, call this the durability choice under monopsony).

This chapter departs from the current focus of the economic literature on durability and posits the expected net present value of the system (Eq. (4.1) discussed in the next section), as the metric that a customer seeks to optimize in his or her choice of the "optimal" durability of a system (think of a customer acquiring a satellite rather than a light bulb).

4.3.3 Value of a System, Optimal Durability Choice, and the Need for a Marginal Cost of Durability Analysis

Why is the analysis of the marginal cost of durability a prerequisite for the choice of durability or design lifetime? The following addresses this question.

To specify the design lifetime requirement, a customer needs to be able to express the present value of a system as a function of its design lifetime. Equation (4.1) is proposed as a means for capturing this value,

$$V(T_{\text{Life}}) = \int_0^{T_{\text{Life}}} [u(t) - c_{\text{OM}}(t)] \times e^{-rt} dt - C(T_{\text{Life}}), \tag{4.1}$$

where

T_{Life}	=	the system's design lifetime,
$V(T_{\text{Life}})$	=	the expected net present value of a system,
$u(t)$	=	the utility rate of a system (e.g., revenues per day for a commercial system),
$c_{\text{OM}}(t)$	=	the cost to operate and maintain the system per unit time,
$C(T_{\text{Life}})$	=	the total system cost prior to fielding or operation, as a function of its durability.

Equation (4.1) is conceptually analogous to the continuity equation (or conservation of mass) in fluid dynamics, which in its integral form is written as follows:

$$\frac{\partial}{\partial t} \int_V \rho dV + \int_S \rho U dS = 0. \tag{4.2}$$

The analogy between the two equations is illustrated in Figure 4.2. The control volume becomes a time interval – the system's design lifetime. The flow entering the control volume is analogous to the aggregate utility or revenues generated during the time bin considered, and the flow exiting the volume corresponds to the cost of acquiring a system designed for this time bin, T_{Life}, plus the cost to operate it during the same period.

Note that Eq. (4.1) can account for the cost of disposal of a system, if any, by appropriately adjusting $c_{\text{OM}}(t)$.

One can now mathematically formulate the question regarding the optimal choice of durability for an engineering system, as seen from the customer's perspective,

$$\begin{cases} V(T_{\text{Life}}) = \int_0^{T_{\text{Life}}} [u(t) - c_{\text{OM}}(t)] \times e^{-rt} dt - C(T_{\text{Life}}), \\ \text{Is there a } T_{\text{Life}}^* \text{ such that } V(T_{\text{Life}}^*) > V(T_{\text{Life}}) \text{ for all } T_{\text{Life}} \neq T_{\text{Life}}^*? \end{cases} \tag{4.3}$$

where T_{Life}^* is the optimal durability of the system.

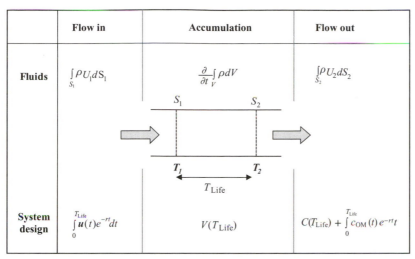

	Flow in	Accumulation	Flow out
Fluids	$\int_{S_1} P U_1 dS_1$	$\frac{\partial}{\partial t} \int_V \rho dV$	$\int_{S_2} P U_2 dS_2$
System design	$\int_0^{T_{\text{Life}}} u(t) e^{-rt} dt$	$V(T_{\text{Life}})$	$C(T_{\text{Life}}) + \int_0^{T_{\text{Life}}} c_{\text{OM}}(t) e^{-rt} t$

Figure 4.2. Analogy between the expected present value of a system as a function of its design lifetime and the continuity equation in fluid dynamics.

It can be seen in Eqs. (4.1) and (4.3) that the analysis of the marginal cost of durability, $C(T_{\text{Life}})$, is a prerequisite for capturing the expected net present value of a system and is therefore a necessary analysis for identifying and selecting the optimal durability of the system.[3]

4.4 Scaling Effects and Marginal Cost of Durability: The Example of a Satellite

The details of the marginal cost of durability are system-specific. This section considers for illustrative purposes the case of a satellite in geostationary orbit and investigates how the different subsystems in the spacecraft scale as a function of its design lifetime requirement. The various effects are then integrated and translated into the marginal cost of durability of the spacecraft. This example is meant to illustrate the essence of the marginal cost of durability analysis.

---

[3] The other multidisciplinary analyses required to identify the optimal durability choice include (1) market analyses and forecast of system expected utility or revenue model $u(t)$, (2) technical analysis and estimate of the cost to operate and maintain the system $c_{\text{OM}}(t)$, and (3) financial analysis of the investment risk in the system, usually referred to as β, which in turn is used to derive the appropriate risk-adjusted discount rate for the investment, r. Each of these analyses raises an interesting set of questions and challenges.

A spacecraft is both an interesting and a paradoxically easy example to illustrate the development of a marginal cost of durability for an engineering system. It is interesting because a spacecraft is composed of many subsystems, for example, the electrical (solar panels, batteries, power control, etc.), thermal (heat pipes, louvers, radiator, etc.), propulsion (propellant, tanks, thrusters, etc.), and attitude control subsystems. Each is impacted differently by the durability or design lifetime requirement. Unlike the case of highway durability, which is mainly controlled by the pavement thickness, as discussed previously, a spacecraft will require many different engineering choices to achieve a required durability, because of the variety of the subsystems involved. The details of these choices will be discussed shortly. On the other hand, a spacecraft is an easy example because, unlike a highway, an aircraft, or a ship, a satellite in geostationary orbit is not physically accessible and therefore cannot be maintained, upgraded, or even replenished (hence, once the propellant is depleted, the spacecraft is out of commission). There is therefore a stronger mapping for a satellite between the initial design choices and the built-in durability than with physically accessible and maintainable systems.[4]

4.4.1 Durability and Spacecraft Subsystems: A Qualitative Discussion

This subsection examines how different spacecraft subsystems scale with the durability or design lifetime requirement. This requirement is a key parameter in sizing several spacecraft subsystems: it directly impacts the design and sizing of some subsystems, for example, the electrical power subsystem, and indirectly impinges on others, for example, the structure, as will be discussed shortly. These influences and coupling are qualitatively captured in Table 4.1. The diagonal in Table 4.1 represents the direct impact of the design lifetime requirement on each subsystem. The off-diagonal terms read as follows: subsystems in the first column scale with the design lifetime,

---

[4] Satellite on-orbit servicing remains a very limited capability confined to low Earth orbit (Saleh, 2003).

Table 4.1. *Durability influence matrix*

	ADCS	TT&C	EPS	Thermal	Structure	Propulsion	Propellant
ADCS[a]	Radiation damage. Shielding / hardening, redundancy		+++	+	+++	+	++
TT&C[b]							
EPS[c]			Solar array degradation, batteries' DOD				
Thermal			+++	Degradation of thermal properties of coating			
Structure	+	+	+++	+		+	++
Propulsion	+					Wear and tear/ on–off cycles	
Propellant	+	+	+++	+	+++	+	Increase in ΔV with design lifetime

[a] Attitude determination and control subsystem.
[b] Telemetry, tracking, and control subsystem.
[c] Electrical power subsystem.

driven by changes in subsystems in the first row. The number of crosses represents the degree of influence (+ + + major influence, + minor influence).

For example, the diagonal term for the telemetry, tracking, and control (TT&C) subsystem[5] reads as follows: increased durability implies longer exposure to space radiation,[6] and therefore a higher probability of failure of electronic components. The best way of protecting against the cumulative effects of radiation damage is by increasing the shielding or hardening of the equipment, which is an expensive procedure. Increased durability therefore increases the cost of the TT&C subsystem through the influence of the increased shielding thickness required to absorb the additional radiation dose. This is the point of the diagonal term for the TT&C subsystem in Table 4.1.

Now consider an off-diagonal term, for example, the thermal subsystem in the first column and the electrical power subsystem (EPS) in the first row; this term will be referred to as the (5; 4) element in the matrix. Table 4.1 indicates a major influence of the latter on the former. What is a spacecraft thermal subsystem? And how is its design influenced by the durability requirement and by the design of the EPS? A spacecraft contains many components that will function properly only if they are maintained within specified temperature ranges. The thermal design of a spacecraft involves identifying the sources of heat, designing proper heat transfer mechanisms between all spacecraft elements, and rejecting heat so that different components stay within their operating temperature ranges (Wertz and Larson, 1999). What matters for the purposes of this section is how the thermal subsystem scales with the durability requirement. A spacecraft is a coupled design, and changes in one subsystem often necessitate changes in another subsystem. The off-diagonal terms in Table 4.1 identify the magnitude of the indirect

---

[5] The TT&C subsystem interfaces between the spacecraft and the ground segment. This subsystem provides the hardware required for the reception, processing, storing, multiplexing, and transmission of satellite telemetry data.

[6] Space radiation consists of (1) trapped radiations in the Van Allen belts in low Earth orbit – electrons, protons, and to a lesser extend O^+ and other heavy ions; (2) galactic cosmic rays (GCRs) – high- to very-high-energy particles (1–1000 MeV), protons, and heavy ionized nuclei from interplanetary sources; and (3) radiation from solar wind and solar flare events (Hastings and Garrett, 1996).

influence that the durability requirement has on one spacecraft subsystem through its impact on another subsystem. For example, increased spacecraft durability necessitates larger solar arrays and batteries, as will be discussed shortly. This overdesign of the EPS has a major influence on the spacecraft thermal subsystem, which has to handle the excess power generation, storage, and distribution. In short, the spacecraft durability requirement impacts the sizing of the thermal subsystem, which in turn impacts the sizing of the thermal subsystem.

The following subsection discusses the quantitative details of the impact of durability on the design of the EPS and the propulsion subsystems.

4.4.2 Durability, the Electrical Power Subsystem, and the Spacecraft Solar Panels

The electrical power subsystem generates power, conditions and regulates it, stores it for peak demand or eclipse operation, and distributes it throughout the spacecraft (Wertz and Larson, 1999). Durability is a key parameter in sizing the EPS. It directly impacts (1) the life degradation of the solar arrays, hence their surface, and consequently their mass, and (2) the battery capacity through the extended number of cycles, hence the reduced depth of discharge (DOD) with design life.

Life degradation of solar arrays is a function of the design lifetime. It occurs for a number of reasons, such as radiation damage and thermal cycling in and out of eclipse, and is estimated as follows:

$$L_d = (1 - \deg radation\ year)^{T_{Life}}. \tag{4.4}$$

The degradation per year is a function of the spacecraft orbital parameters (e.g., position with respect to the Van Allen belts) as well as the solar cycle. A typical value for the degradation factor is 1.2% per year for gallium arsenide (GaAs) cells in geostationary orbit. In low Earth orbit, life degradation can be as high as 2.75% per year (Wertz and Larson, 1999). The solar array's performance at the end of life (EOL), compared with what it was at the beginning of life (BOL), is given by

$$P_{EOL} = P_{BOL} \times L_d. \tag{4.5}$$

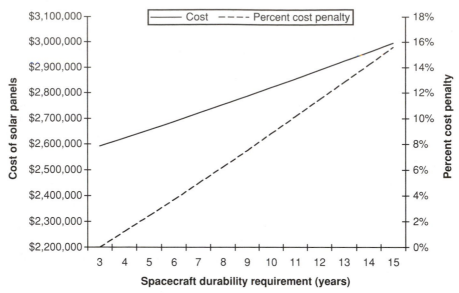

Figure 4.3. Marginal cost of durability of the solar panels of a satellite in geostationary orbit (assumptions: $P_{EOL} = 5$ kW; GaAs cells; specific cost $500/W).

Given a power requirement at EOL, the power output of the solar arrays at BOL scales inversely with life degradation, and the solar arrays have to be overdesigned to accommodate this performance degradation. This overdesign of the solar arrays translates into mass and cost penalties as the durability requirement for the satellite increases. Assuming a power requirement $P_{EOL} = 5$ kW, and a specific cost of the solar arrays of $500/W (Hyder et al., 2000), one can calculate the marginal cost of durability of the satellite's solar panels. The results are shown in Figure 4.3. For the satellite example in geostationary orbit here considered, the marginal cost of durability of the solar panels is approximately a 1.3% increase per additional year of durability requirement.

4.4.3 Durability and the Spacecraft Batteries

Satellite batteries are another interesting example to illustrate the impact of the durability requirement and conduct a marginal-cost-of-durability analysis.

Spacecraft in Earth orbit undergo between 90 and 5500 eclipses per year. The former figure is typical of a satellite in geostationary orbit, the latter of a satellite in low Earth orbit. During eclipse, electric power is supplied by the batteries, which are recharged by the solar arrays when the spacecraft re-emerges into sunlight. In addition, there are some instances when batteries are called upon to provide peak power in sunlight periods. Space-qualified satellite batteries such as nickel–hydrogen are heavy and can constitute up to 15% of the dry mass of a typical satellite (Dudley and Verniolle, 1997).

How does the satellite durability requirement impact the sizing of the batteries? The story may appear a little bit convoluted, but it need not be, and it goes as follows. The amount of energy available from the batteries, which is captured by a (normalized) parameter referred to as depth of discharge, decreases with the number of cycles of charging and discharging. To first order, the number of charge/discharge cycles is equal to the number of eclipses a satellite undergoes during its design lifetime. Typically, a satellite in GEO undergoes two periods of 45 days per year with eclipses, hence 90 cycles of charging and discharging per year. As the satellite durability requirement increases, the number of charging/discharging cycles a battery has to undergo increases; therefore its depth of discharge decreases. Battery capacity (size, mass, and cost) scales inversely with the DOD; therefore, as the spacecraft durability increases, batteries have to be overdesigned to compensate for the reduction in DOD. This again results in a mass and cost penalty for the spacecraft as its design lifetime increases:

$$C_r = \frac{P_e \times T_e}{(DOD) \times N \times n},$$ (4.6)

where

P_e = Power requirement during eclipse (W),
T_e = Duration of eclipse (h),
N = Number of batteries,
n = Transmission efficiency between the batteries and the load, typically 90%.

Figure 4.4 shows the change in DOD for two battery technologies as a function of the number of charge and discharge cycles. For example, for the

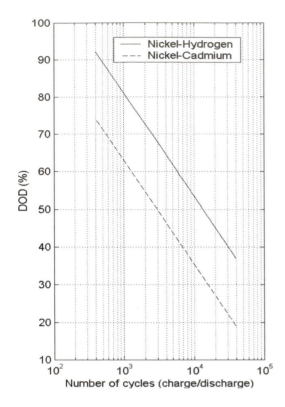

Figure 4.4. Batteries' depth of discharge as a function of charge and discharge cycles.

newer, currently most used nickel–hydrogen technology, a satellite designed with a 5-year durability requirement can extract 90% of the battery capacity (DOD), knowing it will undergo approximately 450 charge and discharge cycles, whereas a satellite with a 15-year durability requirement can extract only 65% of the battery's capacity. Equation (4.6) shows that a decrease in DOD implies the need for a larger capacity, and consequently a more expensive battery.

4.4.4 Durability and the Spacecraft Propellant Budget

As in the previous sections, the interest here is in how the satellite propellant scales with the satellite design lifetime. Propellant is required on board a spacecraft for a number of reasons, for example, to change the spacecraft orbital parameters (e.g., orbit transfer), to correct for errors due to dispersion injection of the launcher, or to counter disturbance forces (e.g., drag in low Earth orbit or third-body gravitational attraction in GEO). The propellant budget also includes a provision for end-of-life disposal of the satellite, for example, to raise the altitude of the orbit if the spacecraft is in GEO (35,786 km) to a junkyard orbit 100–150 km above GEO.

The total velocity change ΔV_{tot} is converted to propellant mass as follows:

$$M_p = M_0 \left[1 - e^{-\left(\frac{\Delta V}{I_{sp}g}\right)} \right], \qquad (4.7)$$

where

g	=	gravitational field strength at sea level (m/s^2),
I_{sp}	=	specific impulse of the propulsion system (s),
M_0	=	initial spacecraft mass or mass prior to the maneuver (kg),
M_p	=	mass of propellant required for a given velocity increment (kg),
ΔV	=	velocity increment for the orbit change or maintenance (m/s).

The detailed derivations of the following need not concern us here, but typically for a satellite in geostationary orbit, a ΔV_{yr} of approximately 50 m/s per year is required for station-keeping (i.e., not letting the satellite drift outside its allowable orbital window). Therefore, the propellant required for station-keeping for T_{Life} years of durability for a spacecraft in GEO is given by

$$\Delta V_{stk} = T_{life} \times \Delta V_{yr}. \qquad (4.8)$$

It is approximately 500 m/s for a 10-year spacecraft and 750 m/s for a 15-year spacecraft. This ΔV increase with spacecraft durability translates into a mass increase through Eq. (4.7). Table 4.2 provides the propellant mass increase for 10-year and 15-year durability for station-keeping of a satellite in geostationary orbit. An initial total mass of 2000 kg is assumed for the satellite.

The propellant mass increase, in turn, requires a larger tank, and both effects – more propellant and a larger tank – translate into an increased cost of the spacecraft propulsion subsystem as the durability requirement increases.

One difficulty should be noted, however, in analyzing the marginal cost of durability of the spacecraft propulsion subsystem. To calculate the increase in propellant mass in Table 4.2, an initial spacecraft mass was

Table 4.2. *Propellant mass for different technologies required for GEO station-keeping (assuming a satellite with an initial total mass of 2000 kg)*

Propulsion technology	$I_{\text{sp}}(s)$	Propellant for station-keeping for 10 years in GEO (kg)	Propellant for station-keeping for 15 years in GEO (kg)
Liquid propellant (hydrazine)	~320	294	425
Electric propulsion (Hall effect thrusters)	~2000	50	75

first assumed, and this number was used to transform our ΔV requirement into a propellant mass requirement. However, as can be seen in Table 4.1, the spacecraft propulsion subsystem is highly influenced by all the other subsystems on board a spacecraft, and its sizing cannot be calculated independently of them. For example, as the durability increases, the power subsystem of the spacecraft becomes more massive (it can weigh up to 15% of a spacecraft mass). This additional mass will require additional propellant to perform a maneuver with a required ΔV, as seen in Eq. (4.7).

A spacecraft durability requirement therefore impacts the propellant budget in two ways: the first is through its direct impact on the total ΔV required, as shown in Eq. (4.8), and the second is through its impact on the other subsystems, which contribute to increase the spacecraft mass, and which in turn necessitate additional propellant to maneuver the spacecraft – the propellant mass needed to provide a required ΔV is a linear function of M_0, as shown in Eq. (4.7). In short, the marginal cost ($ and mass) of durability of the propulsion subsystem cannot be conducted independent of the other subsystems of a spacecraft. The analytical details of the analysis are a little bit involved and would constitute a diversion from the focus of this chapter; they are thus omitted. The interested reader is referred to Saleh et al. (2002) for a detailed discussion of the subject.

4.4.5 Results of the Marginal Cost of Durability of a Satellite

When one aggregates the direct and indirect impact of the durability requirement on all subsystems, spacecraft mass and cost profiles as a function of

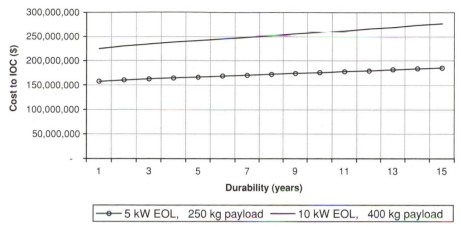

Figure 4.5. Spacecraft marginal cost of durability (spacecraft in GEO, mission reliability = 95%, GaAs cells, Ni–H$_2$ batteries). A satellite's cost to orbit is referred to as the cost to initial operational capability (IOC).

the durability requirement are generated. These mass and cost profiles[7] are the outcome of the marginal cost of durability of a satellite. They need not be continuous and can be calculated for only a few durability points (e.g., 8, 10, 12, and 14 years). The typical marginal cost of durability for a satellite in geostationary orbit is shown in Figure 4.5. A cost penalty, for example, on the order of 10% is in carried in designing such a spacecraft for 15 years as opposed to 10 years. The marginal cost of durability of such satellites can vary between 1% and 5% increase per additional year of the durability and requirement.

   Figure 4.5 provides an illustration of the old adage, although from a different angle than that initially intended, that time is money; that is, more time (durability) requires more resources to develop and embed in the system.

4.5 Cost Elasticity of Durability

This section builds on the previous discussion of the marginal cost of durability to introduce the closely related concept of the cost elasticity of durability.

---

[7] Mass is included because, for satellites, launch cost is a major component (roughly 40% of total cost), and the choice of the launcher depends on the satellite mass. More capable launchers are usually more expensive.

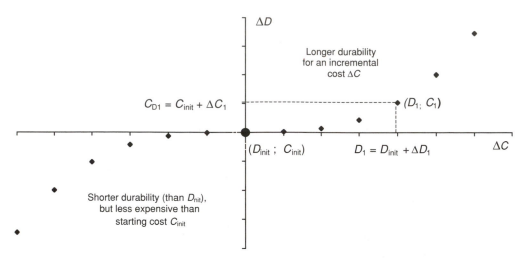

Figure 4.6. The "cost–durability" characterization of components or systems.

It is assumed that the system designers have developed a series of point designs with different durabilities and assessed their total cost. It is also assumed for simplification that these different systems have similar performance and that their only differentiating attributes are durability and cost. These various systems can now be "located" on a cost–durability plot as illustrated in Figure 4.6 (the analysis can also be conducted at the component or subsystem level).

Each point in Figure 4.6 represents a different design. An initial cost–durability pair $(D_{init}; C_{init})$ is chosen to serve as the origin of the plot. The four quadrants on the plot read as follows: the lower-left quadrant consists of systems (or components) that are both less durable and less expensive than the "initial" choice; the lower-right quadrant consists of systems (or components) that are less durable and more expensive than the "initial" choice – therefore they can already be discarded in any further consideration; the upper-right quadrant consists of systems (or components) that are both more durable and more expensive than the "initial" choice; the upper-left quadrant consists of systems (or components) that are more durable and less expensive than the initial choice – this quadrant can remain empty if one chooses the initial design appropriately to serve as reference for the plot.

Figure 4.6 is used for illustrative purposes only, and the various cost–durability pairs need not represent the curve shown. Different curves with different slopes can be obtained (even though technically there are no curves or slopes per se on the plot). In the example used, however, one can read from the plot that large incremental costs around the initial design (D_{init}; C_{init}) result in minor durability extension (see the slope at the origin in Figure 4.6). This observation leads one to introduce a general metric for measuring the sensitivity of different designs' durability to cost perturbations; this metric is termed the *cost elasticity of durability*, by analogy with the price elasticity of demand or supply in microeconomics, and is defined as follows:

$$\varepsilon_{D,C} = \left(\frac{\Delta D}{D} \Big/ \frac{\Delta C}{C} \right). \tag{4.9}$$

This metric measures the relative changes in durability that can be obtained for a given (relative) change in cost. For example, when $\varepsilon_{D,C}$ is large (elastic), one can interpret the result to mean that small changes in cost get us much more durable components or systems. Conversely, when $\varepsilon_{D,C}$ is small (inelastic), changes in cost can buy us only minor durability increments. In the example in Figure 4.6, the initial design with (D_{init}; C_{init}) is highly inelastic: its cost elasticity of durability at C_{init} is close to zero.

Recall that this chapter focuses on the marginal cost of durability in order to identify and select the optimal durability for an engineering system. Equation (4.3) provides the framework for searching for the global optimal durability. But consider, for example, that, instead of the system cost profile $C(T_{\text{Life}})$, there are only two point designs available (shown in Figure 4.6) with two different durabilities, (D_{init}; C_{init}) and ($D_{\text{init}} + \Delta D_1$; $C_{\text{init}} + \Delta C_1$). How does one identify which durability is more appropriate for the system, from a value standpoint?

Instead of using Eq. (4.3), one can compare the incremental present value of the system provided by the incremental durability (ΔD) with the incremental cost (ΔC) necessary to achieve this life extension. Equation (4.10) can be used to assess whether this incremental durability provides a positive incremental NPV or not (and thus whether it should be accepted or rejected):

$$\Delta \text{NPV}_{\Delta D} = \Delta \text{PV}_{\Delta D} - \Delta C_{\text{total}}. \tag{4.10}$$

4.6 From Marginal Cost of Durability to Cost per Day: Regions and Archetypes

Having determined a marginal cost of durability for a system, one can define a cost per day as the ratio of the total system cost to its durability requirement:

$$C_{\text{per\_day}} = \frac{C(T_{\text{Life}})}{T_{\text{Life}}(\text{days})}. \tag{4.11}$$

This definition corresponds to uniformly amortizing the cost of a system over its intended design lifetime. The cost-per-day metric allows a better visualization of the *economies of scale*, if any, that result from extended durability than the marginal cost of durability. How is that?

The marginal cost of durability analysis provides an answer to how capital-intensive durability is for a particular artifact. Economies of scale (with respect to durability), however, are a measure of how fast the system's cost scales compared with the incremental durability provided for the additional cost.

The cost-per-day metric is a good proxy for these economies of scale of durability, because it is a normalized measure of cost per unit time. For example, for the satellite considered in Section 4.4, the cost per day is given in Figure 4.7. It is worth $69,000 per day for a 10-year satellite and $50,000 per year for a 15-year satellite. There are $19,000 economies of scale per unit time for the longer-lived satellite (the larger payload satellite in Figure 4.7).

The spacecraft cost per day decreases monotonically with durability. *In the absence of other metrics*, this behavior of the cost per day may provide a justification for designers to always push the technical boundary and seek an increasingly longer durability. Figure 4.7 also suggests that a customer may be better off requesting the satellite manufacturer to provide the maximum achievable durability, given the economies of scale achieved through extended durability. This argument, however, is flawed and breaks down when one factors in the time value of money and the risk of system obsolescence (e.g., selecting the smallest possible cost per day does not necessarily result in maximizing the system's value). Challenging this cost-centric mindset is discussed in the following chapter.

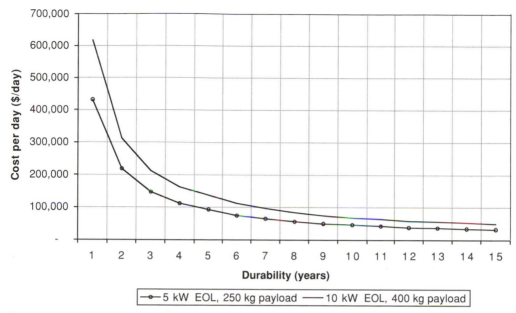

Figure 4.7. Spacecraft cost per operational day (spacecraft in GEO, mission reliability = 95%, GaAs cells, Ni–H$_2$ batteries).

4.6.1 The Durability and Cost per Day Space for an Artifact

More generally, one can define different regions in the durability–cost per day space, as shown in Figure 4.8. The boundaries of each region are considered first.

Zero marginal cost of durability: The lower boundary in Figure 4.8 is defined by the zero marginal-cost-of-durability case. This boundary corresponds to an artifact whose total cost prior to fielding is independent of its durability, or, said differently, the system's cost is insensitive to and does not scale up with increased durability. This boundary is defined by

$$\frac{\partial C(T_{\text{Life}})}{\partial T_{\text{Life}}} \equiv 0. \tag{4.12}$$

It is worth noting than many models in the current economic literature on durability assume infinitely durable goods, and therefore a zero marginal cost of durability (Waldman, 2003). Figure 4.8 shows how limited and particular this assumption is. In addition, designers and engineers may find this assumption highly unrealistic.

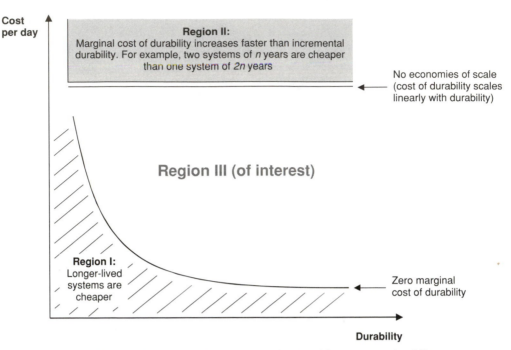

Figure 4.8. Cost per day and economies of scale with respect to durability.

No economies of scale: The upper boundary in Figure 4.8 corresponds to an artifact whose total cost prior to fielding scales linearly with durability. In this case, an artifact with a 2-year durability is strictly equivalent on a service cost per unit time basis to an artifact with a 4-year durability. This boundary is defined by

$$C(T_{\text{Life}}) \alpha C_0 \times T_{\text{Life}}$$

or

$$\frac{\partial C(T_{\text{Life}})}{\partial T_{\text{Life}}} \text{ constant for } T_{\text{Life}} > T. \qquad (4.13)$$

4.6.2 Regions and Archetypes in the Durability and Cost per Day Space for an Artifact

Having defined the relevant boundaries in the durability and cost per day space, one can discuss the different resulting regions shown in Figure 4.8.

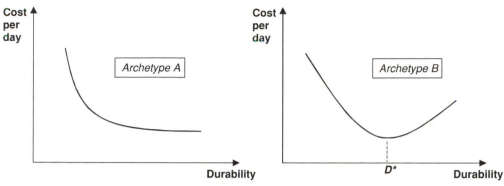

Figure 4.9. Two archetypes for the cost per day.

Region I: This region corresponds to artifacts whose total cost prior to fielding scales down with increased durability. In other words, it is cheaper to build a longer-lived artifact than a shorter-lived one. The problem of durability choice discussed in Sections 4.1 and 4.3 of this chapter is trivial in this region. It is doubtful, however, that, for a given technology, a real design can actually fall into this region.

Region II: This region corresponds to artifacts whose cost scales up faster than the resulting incremental durability. In other words, durability is very capital-intensive in this region (it is expensive to build into the product). For example, two artifacts with 2-year durability are cheaper than one artifact with a 4-year durability in this region. As in the previous case, the durability choice problem is also trivial in this region.

Region III: This region corresponds to artifacts whose total cost prior to fielding scales up with durability, but that also exhibit economies of scale (with respect to durability and cost). Unlike the two previous regions, this region offers an interesting (nontrivial) durability choice problem. Within this region, one can conceive of two archetypes or patterns for an artifact's cost per day (as a function of durability): the first one is a monotonically decreasing curve, and the second one is a curve that exhibits a minimum. These two archetypes are shown in Figure 4.9. The first archetype corresponds to what was shown previously in Section 4.4 with the satellite example of monotonically decreasing cost with durability. The second archetype corresponds to an artifact whose cost, beyond a given durability, rises very steeply. In other words, although there are initial economies of scale, it becomes very expensive to design a longer-lived artifact beyond a given durability.

One should resist the temptation of easy answers with respect to the durability choice problem for these two archetypes. As mentioned previously, archetype A may suggest that the optimal choice of durability, from a customer's perspective, is always the longest durability technically achievable. And archetype B more strongly suggests that the optimal choice of durability is D^*, shown in Figure 4.9. Both deductions, however, are flawed, as the smallest cost per day for a system does not necessarily result in maximizing its net present value. A quantitative justification of this observation is discussed in the following chapter.

4.7 Conclusions

This chapter makes the case that the analysis of the marginal cost of durability is a prerequisite for addressing the durability choice problem of a durable good. In other words, understanding how capital-intensive durability is for a given artifact is a necessary first step toward identifying the optimal durability that maximizes the expected net present value of that artifact. This observation, although straightforward, is often forgotten in the economic literature on durable goods, or dismissed by assuming a zero marginal cost of durability (this assumption is very particular and unrealistic in most cases).

Significant attention has been given to the durability choice problem in the economics literature. Economists, however, have been mainly interested in the choice of durability that maximizes the profits of the manufacturer under monopolistic or competitive market conditions. This chapter departs from the prevalent manufacturer-centric perspective on durability among economists and instead adopts a customer-centric perspective (to make it formal, call this durability choice under monopsony). Furthermore, this chapter posits that, in acquiring a capital good, a customer seeks to maximize the expected net present value of the system in his or her choice of the "optimal" durability (i.e., the system's NPV is the metric to be optimized).

This chapter more specifically focuses on the analysis of the marginal cost of durability of complex engineering products and systems, as opposed to consumer goods (which have been the main focus of economists to date).

Increased product durability, like increased reliability, often requires additional resources and therefore comes at a cost. The relationship between incremental product durability and the incremental cost required to achieve it, if any, is referred to as the analysis of the marginal cost of durability. This analysis is a combination of engineering and cost estimate analyses.

The details of marginal cost of durability are system-specific. To illustrate the application of such an analysis, a satellite example was provided, and the impacts of the durability requirement on the different subsystems on board the spacecraft were quantitatively explored (that is, how the different subsystems, e.g., power, propulsion, and thermal, scale with increased durability). The various effects of durability on the different subsystems were then integrated and satellite cost profiles were provided as a function of durability. The example illustrated the essence and provided a quantitative application of the marginal cost of durability analysis.

Building on the marginal cost of durability analysis, this chapter introduced two closely related metrics, the cost elasticity of durability and the cost per day of an engineering artifact. The cost elasticity of durability metric measures the relative changes in durability that can be obtained for a given (relative) change in cost. Similarly, the cost per day metric was introduced and defined as the ratio of the total system cost to its durability requirement. This metric allows a clear visualization of the *economies of scale*, if any, that result from extended durability.

The metrics and analyses introduced in this chapter contribute a necessary first step toward a rational choice of durability for engineering systems from a customer's perspective. Future work will integrate obsolescence considerations and tradeoffs between capital and maintenance cost for the identification of optimal durability of capital goods.

This chapter is based on an article written by the author and published in the *Journal of Engineering Design*. Used with permission.

REFERENCES

Chamberlin, E. H. "The product as an economic variable." *Quarterly Journal of Economics*, 1953, 67 (1), pp. 1–29.

Dudley, G., and Verniolle, J. "Secondary lithium batteries for spacecraft." *ESA Bulletin No.* 90, May 1997.

Hastings, D., and Garrett, H. *Spacecraft Environment Interactions.* Cambridge University Press, Cambridge, 1996.

Hyder, A. K., Wiley, R. L., Halpert, G., Flood, D. J., and Sabripour, S. *Spacecraft Power Technologies.* Imperial College Press, London, 2000.

Saleh, J. H. "Perspectives in design: The deacon's masterpiece and the hundred-year aircraft, spacecraft, and other complex engineering systems." *Journal of Mechanical Design*, 2005, 127(5), pp. 845–50.

Saleh, J. H., Hastings, D., and Newman, D. "Spacecraft design lifetime." *Journal of Spacecraft and Rockets*, 2002, 39(2), pp. 244–57.

Small, K. A., and Winston, C. "Optimal highway durability." *American Economic Review*, 1988, 78(3), pp. 560–69.

Waldman, M. "Durable goods theory for real world markets." *Journal of Economic Perspectives*, 2003, 17(1), pp. 131–54.

Wertz, R., and Larson, W. *Space Mission Analysis and Design*, 3rd ed. Microcosm Press, Torrence, CA, 1999; Kluwer Academic, Dordrecht, Boston, London.

Wicksell, K. *Lectures on Political Economy.* Routledge and Kegan Paul, London, 1934.

5 Flawed Metrics[1]

System Cost per Day and Cost per Payload

PREVIEW AND GUIDE TO THE CHAPTER

In engineering design, metrics play a critical role in guiding design choices. It is therefore of prime importance that the metrics used in the decision-making process be the "right" metrics. Systems engineers and program managers often invoke the traditional "economies of scale" argument, and the associated metrics, to justify the design of larger, more capable, and longer-lived systems. This chapter challenges the traditional economies of scale argument in system design and makes the case that two metrics used to guide design choices, the *cost per day* and *cost per payload* (or capability), are flawed under certain environmental conditions and result in design choices – increasingly longer-lived systems and larger payloads – that do not necessarily maximize the system's value.

First, this chapter advocates a value-centric mindset in system design and proposes shifting the emphasis from cost to value analyses to guide design choices that maximize a system's value. The case is made that dynamic environmental conditions or competitive markets require a value-centric mindset that views an engineering system as a value-delivery artifact and integrates considerations about the system's cost, its technical environment, and the environment it is serving in order to make appropriate system design choices.

Second, this chapter shows that, although the *cost-per-day* and *cost per payload* metrics are useful guides for design choices in a supply-constrained market (in which a cost-centric mindset can prevail), they are flawed metrics on which to base design choices if the system's environment is uncertain or dynamic or the market is competitive.

---

[1] For guiding design decisions. A metric is never intrinsically flawed; it is only the context of what it is used for that makes it appropriate or flawed.

Because the details of the argument are system- and industry-specific, a satellite example is considered throughout the chapter both for illustrative purposes and in order to avoid abstract exposition and involved theoretic developments. Specifically, it is shown that the *cost per day* and *cost per transponder* are flawed metrics in the sense that they result in design choices – increasingly longer-lived satellites and larger payloads – that do not necessarily maximize the system's value. Counterexamples are provided that challenge the traditional wisdom that longer-lived or bigger satellites, being more cost-effective on a per-day or per-transponder basis, are also more profitable or valuable.

5.1 Introduction

The previous chapter highlighted the importance of the concept of a "metric" in human activities in general and engineering design in particular. A metric was defined as a standard of measurement. It can be measured directly or estimated indirectly, qualitatively or quantitatively, or it can be calculated deterministically or probabilistically by combining different measurements. In an engineering context, metrics were shown in the previous chapter to be necessary prerequisites of the concepts of "performance," "feedback," and "optimization."

System optimization, for example, as mentioned previously, hinges on the notion of a metric, or, as is more familiar to the optimization community, on an *objective function* or *cost function*. These metrics or objective functions will guide design choices by comparing how well each design fares on them. It is therefore of prime importance that the metrics used to guide decision-making be the "right" metrics. This chapter makes the case that two of the metrics used to guide some system design choices, namely the *cost per day* and *cost per payload*, are flawed under certain conditions and result in design choices that do not necessarily maximize the system's value or net worth (considered here the "right" metric, as will be discussed later).

Because the details of the argument are system- and industry-specific, a satellite example is considered throughout the chapter both for illustrative purposes and in order to avoid theoretic economic developments without clear and practical engineering implications.

Table 5.1. *Two metrics for communications satellites, and their design implications*

Metric		Design implication
Satellite cost per day	$C_{\text{per\_day}}$	Impacts selection of the satellite design lifetime. Has driven/justified increase in satellite design lifetime (i.e., resulted in longer-lived satellites).
Satellite cost per transponder	C_{Tx}	Impacts size of payload. Has driven/justified more transponders in satellite payload (i.e., resulted in bigger payloads and consequently bigger satellites).

5.2 Two Metrics in Space System Design and Their Implications

In space system design, and more specifically in communications satellite design, two metrics have often been invoked to guide or justify two particular trends in design choices over the years: longer-lived satellites and bigger payloads. These metrics are *satellite cost per day* and *satellite cost per transponder* (Table 5.1); they have traditionally been used as the quantitative justification for the economies of scale argument in space system design.

Satellite cost per day is defined as the ratio of the spacecraft cost to the initial operational capability (which includes purchasing cost, launch cost, and insurance cost) and its design lifetime, expressed in days:

$$C_{\text{per\_day}} = \frac{C(T_{\text{Life}})}{T_{\text{Life}}(\text{days})}. \tag{5.1}$$

This definition corresponds to uniformly amortizing the cost of the satellite – in which the cost to operate it can be included – over its intended design lifetime. A decreasing cost per day reflects economies of scale (along the time dimension). Similarly, satellite cost per transponder[2] is defined as the ratio of the spacecraft cost to initial operational capability and its total number of transponders (or 36-MHz transponder equivalent):

$$C_{\text{Tx}} = \frac{C(T_{\text{Life}})}{\text{Total number Tx}}. \tag{5.2}$$

This definition corresponds to uniformly amortizing the cost of the satellite – in which again the cost to operate it can be included – over its payload size or total number of transponders on board.

---

[2] The transponder is a particular case of a system's payload.

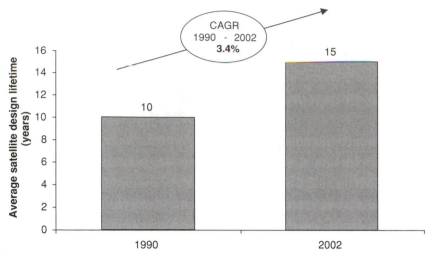

Figure 5.1. Trend in average satellite design lifetime. Compounded annual growth rate 3.4%. *Source:* Futron, 2004.

Figures 5.1 and 5.2 show the trends in average satellite design lifetime and payload size since 1990. The average design lifetime for communications satellites has grown from approximately 10 years in 1990 to 15 years in 2002; and the average number of 36-MHz transponder equivalents on board a spacecraft has grown from 26 in 1990 to 48 in 2002 (at a compounded annual growth rate of 5.2%). It should be noted that these numbers are only averages and some satellite manufacturers, for example, offer bigger and

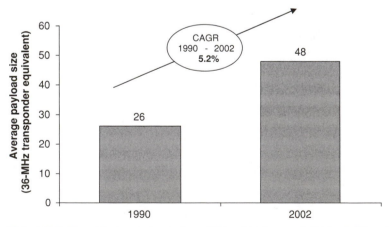

Figure 5.2. Trend in average number of transponders on board satellites. Compounded annual growth rate 5.2%. *Source:* Futron, 2004.

longer-lived satellites (e.g., Boeing B-702, which can carry over 80 transponders, and Astrium Eurostar 3000, which can be fitted to remain operational for 17 years).

The following sections discuss how these two trends map into or are explained by the two previously mentioned metrics. It is then argued that these metrics derive from a cost-centric mindset, and that they are flawed metrics for guiding design choices when the ultimate objective is to maximize the system's value or net worth (not minimize its cost, or some cost-related metric).

5.3 Investigating Satellite Cost per Day

In the previous chapter, a detailed discussion of the marginal cost of durability was presented. Analytical results were derived, and a numerical example was provided in the case of a particular complex engineering system: a communications satellite. It was shown, for example, that a cost penalty of roughly 10% is incurred in designing a satellite for 15 years instead of 10 years. Recall that the analysis of the marginal cost of durability reflects the way the design lifetime requirement impacts the sizing of the different components and modules in a system. More specifically, the marginal cost of durability is the (analytical or numerical) relationship between the incremental design lifetime of a system and the incremental cost required to achieve it. This analysis is a combination of engineering and cost estimate analyses.

The results for the marginal cost of durability and cost per day of a communications satellite are summarized in Figures 5.3 and 5.4 (the details are available in Chapter 4).

Having obtained $C(T_{\text{Life}})$, it is now easy to calculate the satellite cost per day, as given in Eq. (5.1). Figure 5.4 shows typical profiles of this metric. For example, designing a 10-kW communications satellite for 10 years has a cost per day of approximately $69,000 per day, whereas this metric drops to $50,000 per day if the satellite were designed for 15 years ($19,000 advantage on a cost-per-day basis for the longer-lived satellite). The main feature of the behavior of a satellite cost per day is that it decreases monotonically with the satellite design lifetime. In other words, it always costs less on a per-day

86   Flawed Metrics

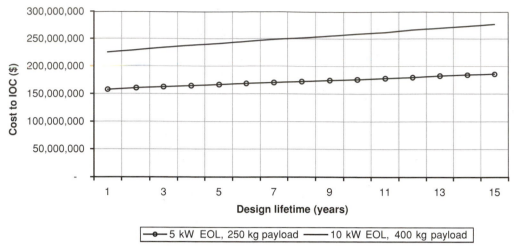

Figure 5.3. Spacecraft $C(T_{Life})$ as a function of the design lifetime requirement (spacecraft in GEO, mission reliability = 95%, GaAs cells, Ni–H$_2$ batteries).

basis to design for a longer lifetime (this is the essence of the economies of scale argument along the time dimension).

It is this result that guided or justified the trend shown in Figure 5.1 of designing and launching longer-lived satellites. In addition, this result

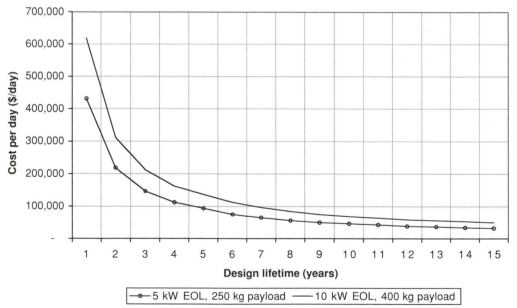

Figure 5.4. Satellite cost per day ($/day) as a function of the design lifetime (same parameters as in Figure 5.3).

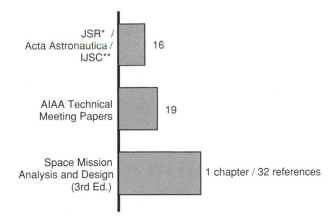

* *Journal of Spacecraft and Rockets*
** *Internation Journal of Satellite Communications and Networking*

Figure 5.5. Proliferation of satellite cost models (as of 02/01/05).

also suggests that customers may always be better off requesting the satellite manufacturer to provide the maximum technically achievable design lifetimes for their satellites:

$$T_{\text{Life-optimal}} = T_{\text{life-max}}. \tag{5.3}$$

All this seems sensible. What can possibly be flawed about the satellite cost-per-day metric and the above findings and design implications? These issues are explored in the following section.

5.4 The Case for a Value-Centric Mindset in System Design

The two previously mentioned metrics derive from a cost-centric mindset, and their calculation is facilitated by the availability of cost models throughout the aerospace industry (see Figure 5.5). The proliferation of cost models is the laudable result of the emphasis over the last two decades on financial discipline in the design of complex engineering systems. The technical literature saw a proliferation of "design to cost" methodologies in the 1980s, the intellectual underpinning of which rests on the development of cost models for hardware and software at the component, subsystem, and system levels. In the space industry, several agencies have developed and refined over the years parametric cost models that relate spacecraft cost or subsystem cost to physical, technical, and performance parameters. A spacecraft's cost, for example, depends on its size, mass, complexity, technology readiness, and design lifetime, as well as other characteristics. Popular spacecraft cost models include the NASA/Air Force Cost Model (NAFCOM) used by NASA, the Air Force, and their respective contractors and the Aerospace Corporation's Small Satellite Cost Model (SSCM). Spacecraft cost models are based on the relevant historical databases of past satellite programs. The basic assumption of parametric cost modeling is that a system or a subsystem will cost the next time what it has cost the previous times.[3] The availability of spacecraft cost models, as mentioned previously, makes it easy to calculate cost-based metrics, such as the cost per day and cost per transponder of a satellite.

Spacecraft cost models are not only popular with industry professionals; they also have made their way into the curricula of aerospace departments at a number of academic institutions in the United States. Students in aerospace departments are being exposed to and educated about cost models, and an increasing number of graduate theses include cost implications among the design tradeoffs that are explored. These students will in turn become aerospace professionals familiar with cost modeling and carry as part of their educational background a sensitivity to cost implications in their technical responsibilities.

So all is well in the best of all possible worlds? Not exactly. Although cost models are pervasive throughout the aerospace industry, quantitative value models of space systems that integrate cost and utility models are quasinonexistent. Why should this be an issue?

Although the proliferation of cost models is laudable, the absence of interest in satellite value or utility models is deplorable. The situation perhaps conveys the false impression that satellites are either cost sinks or expensive artifacts whose value or utility profile over their design lifetime is difficult to quantify and does not warrant efforts to do so. More important, the absence of quantitative value or utility models makes it difficult

---

[3] Stated this way, one can easily see in such an approach both its advantage, of learning from the past, and its limitation, in taking the past as the sole guide for the future (does not allow for breaking the paradigm).

to build a convincing case for such systems to policy makers or decision-makers, especially in the light of their exorbitant costs. Furthermore, the specification and selection of a system design lifetime, or of a system life extension (e.g., the Hubble Space Telescope), will always have weak arguments fraught with subjectivity in the absence of quantitative revenue/utility models.

5.4.1 Satellites as Value-Delivery Artifacts

Satellites, like any other complex engineering systems, should be perceived as value-delivery artifacts. And the value delivered, or the flow of service that the spacecraft delivers over its design lifetime, whether tangible or intangible, deserves as much effort to quantify as the system's cost. The following subsection expands on this proposition and highlights the limitation of the cost- per-day metric in this value-centric context.

5.4.2 Minimizing Cost or Maximizing (Net) Value? Limitations of the Satellite Cost-per-Day Metric

System designers or decision-makers, when invoking results similar to the ones shown in Figure 5.4 to push for increased spacecraft design lifetime, may be using the smaller cost per day as a proxy for higher profitability or return from their investment in the satellite. In other words, the designers' assumption may be that communications satellites with a smaller cost per day (longer-lived satellites) are more profitable than those with a larger cost per day (shorter-lived satellites).

If one posits that the objective of investing in a satellite system is to maximize its profitability (or some other measure of the net worth of the system or its return on investment), then the above assumption is flawed in that it ignores the revenue side in the profitability equation. In simple terms, profitability being the difference between revenues (R) and cost (C), minimizing cost or a cost-related metric does not guarantee maximizing profits (Π),

$$\Pi = R - C. \tag{5.4}$$

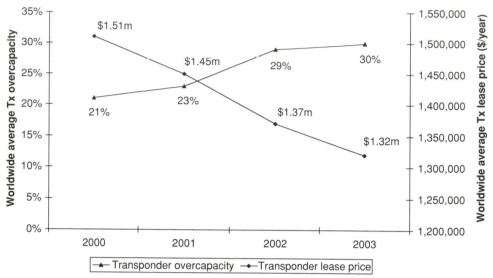

Figure 5.6. Worldwide transponder overcapacity and average transponder lease price. *Data source:* Révillon et al., 2004.

The assumption of minimizing cost to obtain maximum profits or return on investment may be justified in a supply-constrained market[4] in which revenues from the system are guaranteed to remain constant or grow with time. But it is wrong to make this assumption in a competitive market, such as the satellite services market, which currently has a sizable overcapacity of on-orbit transponders (see for example the downward pressure on transponder lease price in Figure 5.6), and in which end customers have significantly more choice and market power to turn away from one provider utilizing "aging" transponders, for example, to another offering newer, better, or more powerful transponders.

What the previous paragraph is suggesting is that in selecting the spacecraft design lifetime, the satellite cost-per-day metric should not be the only one considered. Satellite operators have to assess the risk of loss of value due to both obsolescence of their spacecraft technology base and the likelihood of changing or shifting market needs after the satellite has been launched. For example, it is not obvious that it is in the best interest of a satellite

---

[4] The demand for a given service or product in such a market is not fully met, and there are always customers to purchase any additional product or service that is made available on this market.

operator to make the contract life of a spacecraft too long: markets can shift, and new or enhanced capabilities might be developed and become available within a couple of years following the launch – hence the need to launch a new satellite or risk losing revenues and market share to a competitor who launches later with newer or more advanced capabilities. So how can one capture the value of a system (or the loss of it) as a function of its design lifetime, and can it be shown analytically that minimizing cost (or cost per day) does not correspond to maximizing the system's value?

Equation (5.5) is a more elaborate version of the profitability equation (5.4). It captures the value of an engineering system as a function of its design lifetime:

$$V(T_{\text{Life}}) = \int_0^{T_{\text{Life}}} [u(t) - c_{\text{OM}}(t)] \times e^{-rt} dt - C(T_{\text{Life}}), \quad (5.5)$$

where

T_{Life}	=	system's design lifetime (or durability),
$V(T_{\text{Life}})$	=	expected net present value of a system,
$u(t)$	=	utility rate of a system (e.g., revenues per day for a commercial system),
$c_{\text{OM}}(t)$	=	cost to operate and maintain the system per unit time,
$C(T_{\text{Life}})$	=	total system cost prior to fielding or operation, as a function of durability,
r	=	an appropriate discount rate for the project/system.

Equation (5.5) shows explicitly the functional dependence of the cost of the system on the system's design lifetime (e.g., a spacecraft designed for 7 years will cost less than a spacecraft designed for 15 years).[5]

Consider, for example, the following revenue model (revenue per day) of a satellite in which a time scale of obsolescence affects the system's revenue as follows:

$$u(t) = u_0 \times \exp\left[-\left(\frac{t}{T_{\text{obs}}}\right)^2\right]. \quad (5.6)$$

---

[5] Note that Eq. (5.5) can account for the cost of disposal of a system, if any, by appropriately adjusting $c_{\text{OM}}(t)$.

The model is adapted from the standard component life cycle model, from which the emerging and growth phases are removed (see Chapter 6 for more details). This removal can be partly justified by considering the fact that, although components are mass-produced and their sales gradually ramp up, the systems considered in this chapter reach full operational capability in a length of time that is much shorter than their design lifetime. Other models of obsolescence can of course be used, but will not change the nature of the findings in this section. For illustration purposes, assume the following:

$$
\begin{aligned}
u_0 &= \$80{,}000/\text{day}, \\
r &= 10\%, \\
c_{\text{OM}} &= 7.5\% \text{ of purchasing cost of the satellite}, \\
C_0 &= \$150 \text{ million}, \\
\alpha &= \text{average marginal cost of durability of 4\% increase per year}, \\
C(T_{\text{Life}}) &= C_0(1+\alpha T_{\text{Life}}), \\
T_{\text{obs}} &= 15 \text{ years}.
\end{aligned}
$$

Results for the cost per day as well as the expected net present value of the spacecraft as a function of the design lifetime are shown in Figure 5.7.

Three observations can be made regarding Figure 5.7:

i. First, note on the lower curve with the corresponding y-axis to the right the monotonically decreasing cost per day of a satellite. For example, two points on Figure 5.7 read as follows: for a satellite designed for 8 years versus another designed for 12 years, the cost per day decreases from \$85,000 per day to \$59,000 per day.

ii. Second, note that the expected net present value with the corresponding y-axis to the left has a maximum at around 8 years, after which it decreases. This behavior is due on the one hand to the typical satellite cost profile $C(T_{\text{Life}})$, shown in Figure 5.3, and on the other hand, because of the time value of money and because the forecast revenue profile is modeled by Eq. (5.6) (revenue decreases with time as technology obsolescence effects settle in and drive end-users toward other service providers with newer, better QoS offerings, for example).

iii. Finally, note that given the assumptions, should the designers decide on the 12-year design lifetime instead of 8 years because of the smaller

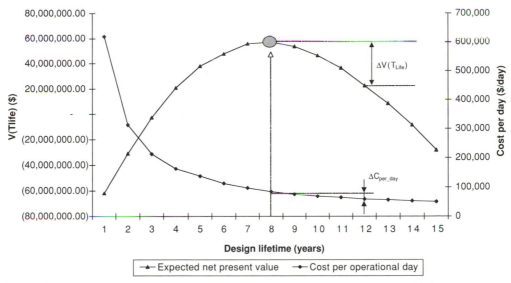

Figure 5.7. Net present value versus cost per day as a function of satellite design lifetime.

cost per day, they would in fact be forfeiting a sizable financial value $\Delta V(T_{\text{Life}})$ by going for the longer-lived satellite, as shown in Figure 5.7. In other words, the decision to select the longer design lifetime because of its smaller cost per day does not correspond to or result in maximizing the value of the system. $\Delta V(T_{\text{Life}})$ can be viewed in this case as a value penalty for selecting the longer-lived satellite.

The purpose of the previous example was to provide one counterexample that challenges the rarely questioned traditional wisdom that longer-lived communications satellites are more cost-effective and therefore more profitable or *valuable*. The reader is referred to Chapter 2 for a discussion of the advantages and disadvantages in reducing versus extending a satellite's design lifetime.

In summary, this section has shown that, although the cost per day metric for a communications satellite may be a useful guide in deciding on a system's design lifetime in a supply-constrained market, it is a flawed metric on which to base this decision if the market is competitive, and the revenues from the system are not guaranteed to remain constant or grow with time. In particular, it was shown that if revenues were modeled by taking into account technology obsolescence effects (e.g., end customers turning away from one provider utilizing "aging" transponders to another offering newer, better, or more powerful transponders), then selecting the longer design lifetime because of its smaller cost per day does not correspond to or result in maximizing the value of the system.

This subsection suggests that, if value (as given in Eq. (5.5)) is taken as the metric to be maximized in the design of communications satellite, then, under certain circumstances (e.g., in competitive markets), shorter-lived satellites are more *valuable* than longer-lived ones.

5.4.3 Beyond the Satellite Example

The previous argument for challenging the cost-per-day metric as a useful guide for design choices in uncertain or dynamic environments rests on the following two assumptions and is valid for any durable (capital) good that satisfies these conditions:

i. **Marginal Cost of Durability:** The asset under consideration has a nonzero marginal cost of durability. In other words, the cost of the asset is not indifferent to its durability and scales up with its design lifetime (i.e., it costs more to design a system for a longer lifetime; see Chapter 4 for more details).

ii. **Network Externalities:** The need for and valuation of the services provided by the asset are driven by endogenous variables and competitive environmental conditions.

5.5 Satellite Cost per Transponder: Design Implications and Limitations

An argument similar to the one advanced previously with respect to the satellite cost per day can be made regarding the satellite cost per transponder (or cost per payload):

i. It is true that bigger satellites, leveraging economies of scale, offer a smaller cost per transponder, as seen in Figure 5.8. In fact, satellite cost per transponder decreases monotonically with the number of transponders. This result may have guided the trend seen in Figure 5.2

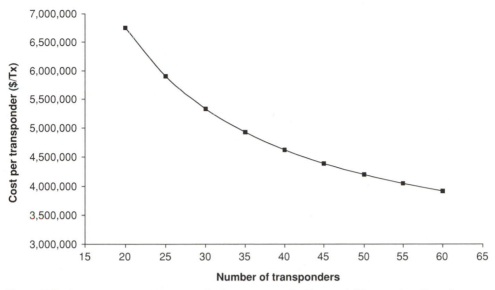

Figure 5.8. Average cost per transponder for a communications satellite as a function of the size of the payload (calculated based on satellite cost models).

of designing increasingly larger communications satellites. Two points, for example, on Figure 5.8 read as follows: a satellite with 40 transponders costs approximately $4.6 million per transponder, whereas a larger satellite with 60 transponders costs approximately $3.9 million per transponder. This represents almost a 20% cost advantage per transponder for the larger satellite.

ii. Smaller cost per transponder may have been used by the industry as a proxy for more economical and therefore more profitable satellites. This assumption, like its analog with respect to smaller cost per day, may be valid in a supply-constrained market, but it is flawed in a competitive market plagued by overcapacity, such as the satellite services market, in which service prices (e.g., transponder lease price) are highly sensitive to the supply–demand imbalance, as will be argued shortly.

Consider the following. The worldwide overcapacity of on-orbit transponders in the year 2000 was 21%. Overcapacity is simply defined here as the percentage of unused transponders (it is the complement of the global utilization rate of GEO communications satellites). Overcapacity has been steadily increasing over the past few years, and by 2003 it had reached 30%

(see Figure 5.6), growing at an annual compounded growth rate of 12.6%. Increasingly bigger satellites (see Figure 5.2) have contributed their share to this growing overcapacity. Unfortunately, as satellites are getting bigger, they are getting less "filled up," as suggested by the decreasing load factor over the past few years – from 79% in 2000 to 71% in 2003 (Révillon et al., 2004). The CEO of Eutelsat recently remarked that "large empty satellites clearly do not contribute to higher profitability" (Berretta, 2003). More important, as satellites get bigger (partly because of the smaller cost per transponder) and overcapacity increases, competition heightens among satellite operations and puts downward pressure on transponder lease prices. For example, the global average transponder lease price in 2000 was about $1.5 million/year (with 21% overcapacity), but it dropped to $1.3 million/year in 2003 (as overcapacity increased to 30%) at an annual compounded growth rate of –4.3%. Figure 5.6 superimposes the evolution of the worldwide overcapacity of on-orbit transponders and the average transponder lease price.

When the effects of bigger satellites and smaller transponder lease prices are aggregated, what is the net outcome? Consider for example the revenues generated per year by the global "average" satellite in 2000 (load factor 79%, transponder lease price $1.51 million/year, and payload of 43 transponders – calculated using the CAGR from Figure 5.2) compared with the revenues generated by the "average" satellite in 2003. The revenues generated per year by a communications satellite can be captured by the following expression:

$$u(t) = N_{\text{Tx\_total}} \times L(t) \times \overline{P}(t), \qquad (5.7)$$

where

$L(t)$	=	Average satellite load factor,
$N_{\text{Tx\_total}}$	=	Total number of transponders on board a satellite,
$\overline{P}(t)$	=	Average transponder lease price per unit time.

The reader is referred to Appendix A for the development and discussion of this expression. Results shown in Table 5.2 suggest that the smaller average satellite in 2000 generated more revenues than its larger counterpart in 2003, given the smaller load factor and depressed transponder lease price in 2003.

Table 5.2. *Comparison of revenues generated by the "average" satellite in 2000 versus 2003*

	Average load factor	Average transponder lease price	Average payload size	Revenues generated in the given year
2000	79%	$1.52 million/year	43 transponders	$52 million
2003	70%	$1.32 million/year	50 transponders	$47 million

This example challenges the traditional wisdom that bigger satellites, being more cost-effective on a per-transponder basis, are also more profitable. They are not necessarily so in a competitive market with significant overcapacity and in which service price is sensitive to supply–demand imbalance. The result above can be more carefully studied in light of game theory: all players (satellite operators), in acquiring larger satellites with increasingly more transponders and therefore smaller cost per transponder in the hope of undercutting the competition, have in fact contributed to the market overcapacity and put downward pressure on transponder lease price. Decreased transponder lease prices in the above example have outweighed the cost advantage on a per-transponder basis on larger satellites.

In summary, this section has shown that, although the cost per transponder metric for a communications satellite may be a useful guide in deciding on the payload size (number of transponders) in a supply-constrained market, it is a flawed metric on which to base this decision if the market is competitive. Careful consideration should be given to overcapacity in the market as well as its potential impact on transponder lease prices. This section also suggests that, in certain markets, smaller satellites may be more profitable than bigger ones.

5.6 Conclusions

This chapter challenges the traditional "economies of scale" argument in system design and makes the case that two metrics associated with this argument, and used to guide design choices, namely cost per day and cost per payload (or capability), are flawed under certain environmental conditions and result in design choices – increasingly longer-lived systems and larger payloads – that do not necessarily maximize the system's value.

Because the details of the argument are system- and industry-specific, a satellite example is considered throughout the chapter. It is shown that two metrics used to guide communications satellites design, namely satellite cost per day and satellite cost per transponder, are flawed under certain conditions and result in design choices – increasingly longer-lived satellites and increasingly larger payloads – that do not necessarily maximize the system's value or net worth.

Sections 5.2 and 5.3 showed that satellite cost per day decreases monotonically with satellite design lifetime. In other words, it always costs less on a per-day basis to design for longer lifetime. It is argued that this result has guided or justified the trend of designing and launching longer-lived satellites. System designers or decision-makers, when invoking this result, may be using the smaller cost per day as a proxy for higher profitability or return from their investment in the satellite. In other words, the designers' assumption may be that communications satellites with smaller cost per day (longer-lived satellites) are more profitable than those with a larger cost per day (shorter-lived satellites).

In Section 5.4, the previous assumption is shown to be flawed in that it ignores the revenue side in the profitability equation. The assumption of minimizing cost to obtain maximum profits or return on investment may be justified in a supply-constrained market in which revenues from the system are guaranteed to remain constant or grow with time. But it is wrong to make this assumption in a competitive market, such as the satellite services market, which currently has a sizable overcapacity of on-orbit transponders, and in which end customers have significantly more choice and market power. A counterexample is provided that challenges the traditional wisdom that longer-lived communications satellites are more cost-effective and therefore more profitable or valuable. In particular, it is shown that, if revenues are modeled by taking into account technology obsolescence effects (e.g., end customers turning away from one provider utilizing "aging" transponders to another offering newer, better, or more powerful transponders), then selecting the longer design lifetime because of its smaller cost per day does not correspond to or result in maximizing the value of the system.

In Section 5.5, it is shown that larger satellites leveraging economies of scale offer a smaller cost per transponder. In other words, it always costs less on a per-transponder basis to have a larger satellite with more transponders on board. However, smaller cost per transponder is not a good proxy for more profitable or *valuable* satellites, as is also shown in Section 5.5. An example is provided that challenges the traditional wisdom that bigger satellites, being more cost-effective on a per transponder basis, are also more profitable. They are not necessarily so in a competitive market with significant overcapacity and in which service price is sensitive to the supply–demand imbalance. It is also shown that, although the cost per transponder metric for a communications satellite may be a useful guide in deciding on the payload size in a supply-constrained market, it is a flawed metric on which to base this decision if the market is competitive. Careful consideration should be given to overcapacity in the market as well as its potential impact on transponder lease prices.

In summary, the two metrics discussed in this chapter derive from a cost-centric mindset, and their calculation is facilitated by the availability of cost models throughout the aerospace industry. These metrics are useful guides for design choices in a supply-constrained market (in which a cost-centric mindset can prevail), but they are flawed metrics on which to base decisions if the market is competitive, and in which the revenues from the system are not guaranteed to remain constant or grow over time, through the impact of technology obsolescence for example, or when the market is plagued with overcapacity, thus putting downward pressure on transponder lease prices. Finally, this chapter makes the case that the current market conditions require a value-centric mindset that views a spacecraft as a value-delivery artifact and integrates considerations about the system's cost and about its technical environment and the market it is serving.

This chapter is based on an article written by the author and published in the *IEEE Transactions on Aerospace and Electronic Systems*.

REFERENCES

Berretta, G. "Strategies for Continued Growth." Paris, September 9, 2003. Available online at http://www.geosat.lu/Download/EutelsatFacts/EUTELSAT_ SEPT2003_PRES.pdf.

Futron. "How many satellites are enough? A forecast of demand for satellites 2004–2012." Futron Corporation, Bethesda, MD, February 2004. Available

online   at   www.futron.com/pdf/resource_center/white_papers/Satellite_
Forecast_2004_2012_White_Paper.pdf (accessed 7/15/07).

Révillon, P., Villain, R., Bochinger, S., Gallula, K., Pechberty, M., Rousier, A., and
Bellin, S. *World Satellite Communications & Broadcasting Markets Survey, Ten
Year Outlook (2004 Edition)*. Euroconsult, Paris, August 2004.

6 Durability Choice and Optimal Design Lifetime for Complex Engineering Systems

PREVIEW AND GUIDE TO THE CHAPTER

This chapter addresses the durability choice problem for complex engineering systems, as seen from the customer's perspective and in the face of network externalities and obsolescence effects. Economists have investigated the impact of market structure on manufacturers' durability choices (under monopolistic or competitive market conditions); this chapter departs from the literature on the subject by addressing the durability choice problem from the customer's perspective, and an "optimal" durability is sought that maximizes the NPV of an asset for the customer (as opposed to maximizing the profits of the manufacturer). First, the various qualitative implications for reducing or extending a product's durability are discussed. Second, analytical results for the optimal durability are derived under steady-state and deterministic assumptions. Trends and functional dependence of the optimal durability on various parameters are identified and discussed. Third, the durability choice problem is explored when the risk of obsolescence is accounted for. Finally, the durability choice problem under uncertainty is investigated and the various risks in making *cautious* or *risky* choices of durability are discussed.

6.1 Introduction: A Topic Overlooked by Economists and Engineers

The objective of this chapter is to contribute an analytical framework toward the rational choice of durability for engineering systems, as seen from a customer's perspective, and in the face of network externalities and obsolescence effects.

The chapter is organized as follows: Section 6.2 argues for an augmented perspective on engineering design and optimization, as a prerequisite for addressing the durability choice problem of complex engineering systems. Section 6.3 derives analytical results for optimal durability under steady-state and deterministic assumptions. Section 6.4 expands these results to the dynamic case and accounts for depreciation and obsolescence in the durability choice problem. Section 6.5 discusses the durability choice problem under uncertainty. Section 6.6 provides a conclusion to this work.

6.2 An Augmented Perspective on Design and Optimization: A System's Value and the Associated Flow of Service

Engineering (system) design is traditionally viewed as a matching between two (vector) quantities: resources and system performance. One traditional design paradigm fixes the amount of available resources and optimizes system performance given the resource constraint. The other approach constrains the system performance to a desired level and strives to find a design that will achieve this performance at minimal cost. The first approach operates with – and attempts to maximize – a performance per unit cost metric; the second approach seeks to minimize a cost per function (or capability) metric. Several powerful optimization techniques have been developed and continue to be successfully applied in engineering disciplines based on these two design paradigms, such as gradient-based or population-based methods (e.g., genetic algorithms). See for example Papalambros and Wilde (2000) or Gen and Cheng (1997) for an overview of different such techniques. However, these paradigms, and the associated metrics, are not sufficient to address issues of design in the time dimension. *To (quantitatively) discuss issues related to durability or design lifetime, it is important to first conceive of a design, not only as a technical achievement, but also as a value-delivery artifact. And the value (to be) delivered or the flow of service that the system would provide over time, whether tangible or intangible, deserves as much effort to quantify as the system's cost.* I refer to this perspective as a value-centric mindset in system design, as opposed to the traditional cost-centric mindset. *The distinction is not only an academic exercise but has direct and practical design implications: different design decisions will ensue when one adopts a cost-centric or a value-centric approach to design,* as will be shown later.

6.2.1 Durability and Net Present Value

From a practical perspective, the previous discussion emphasizes the need to introduce metrics per unit time (e.g., cost and utility) and to express the value of a system as a function of its durability. To select the optimal durability of an engineering system – recall that optimality in this case refers to the durability that maximizes the system's NPV – a customer needs to be able to express the value of a system as a function of its durability. Equation (6.1) offers one way of capturing this value:

$$V(T_{\text{Life}}) = \int_0^{T_{\text{Life}}} [u(t) - c_{\text{OM}}(t)] \times e^{-rt} dt - C(T_{\text{Life}}),\qquad(6.1)$$

where

T_{Life}	=	system's design lifetime (or durability),
$V(T_{\text{Life}})$	=	expected net present value of a system,
$u(t)$	=	utility rate of a system (e.g., revenues per day for a commercial system),
$c_{\text{OM}}(t)$	=	cost to operate and maintain the system per unit time,
$C(T_{\text{Life}})$	=	total system cost prior to fielding or operation, as a function of its durability.

Note that Eq. (6.1) can account for the cost of disposal of a system, if any, by appropriately adjusting $c_{\text{OM}}(t)$.

Given Eq. (6.1), which captures the expected NPV of a system (and shows the functional dependence on the system's durability), the durability choice problem from the customer's perspective can be expressed mathematically as follows:

$$\begin{cases} V(T_{\text{Life}}) = \int_0^{T_{\text{Life}}} [u(t) - c_{\text{OM}}(t)] \times e^{-rt} dt - C(T_{\text{Life}}). \\ \text{Is there a } T_{\text{Life}}^* \text{ such that } V(T_{\text{Life}}^*) > V(T_{\text{Life}}) \text{ for all } T_{\text{Life}} \neq T_{\text{Life}}^*, \end{cases}\qquad(6.2)$$

where T_{Life}^* is the optimal durability of the system.

In words, T_{Life}^* is the durability choice that maximizes the system's NPV.

6.2.2 Quantitative Analyses Required to Address the Durability Choice Problem

Equations (6.1) and (6.2) indicate that four different types of analyses are required for identifying and selecting the optimal durability of the system. These are

1. Marginal cost of durability of the system, $C(T_{\text{Life}})$, or how the total cost of the system prior to fielding scales with durability – in other words, how capital-intensive durability is for the particular system considered.
2. Market analyses and forecasts of system expected utility or revenue model, $u(t)$.
3. Technical analyses and estimates of the cost to operate and maintain the system per unit time, $c_{\text{OM}}(t)$
4. Financial analyses of the investment risk in acquiring the system, usually referred to as β, which in turn is used to derive the appropriate risk-adjusted discount rate for the investment, r.

6.3 Optimal Durability under Steady-State and Deterministic Assumptions

This section examines the simple case of the durability choice problem under steady-state conditions and deterministic assumptions, namely, (1) constant revenues and (2) no technology obsolescence or market volatility effects throughout the intended duration of operation of the system. The purpose of this section is to identify trends and functional dependence of the optimal durability in the various analyses discussed in the previous section. The section makes no claim to numerical accuracy, and realism is sacrificed in this first step in order to facilitate understanding of the problem and avoid analytical complications that could clutter the main focus.

Two other assumptions are made in this section:

3. The cost to operate and maintain the system per unit time is constant throughout the operational life of the system. Notice that this assumption can be easily amended to include a cost profile for operations as a function of the mission phase (e.g., initial fielding, nominal operations,

contingency operations, and aging and deterioration). In addition, the utility derived or revenue generated is larger than the cost to operate and maintain the system.

4. The marginal cost of durability is linear. In other words, the system's cost scales linearly with durability. This assumption can also be easily modified to accommodate better fits and more involved models of the marginal cost of durability.

Given these assumptions, Eq. (6.2) has a simple closed-form solution that is obtained by setting the first-order condition to zero,

$$\left. \frac{\partial V}{\partial T_{\text{Life}}} \right|_{T_{\text{Life}}^*} = 0, \tag{6.3}$$

and verifying that the second-order condition is satisfied (for a maximum),

$$\left. \frac{\partial^2 V}{\partial T_{\text{Life}}^2} \right|_{T_{\text{Life}}^*} = -r \left(u_0 - c_{\text{OM}} \right) e^{-r T_{\text{Life}}} < 0. \tag{6.4}$$

The optimal durability of the asset, under the assumptions discussed at the beginning of this section, is given by Eq. (6.5):

$$T_{\text{Life}}^* = -\frac{1}{r} \ln \left(\frac{C_0 \alpha}{u_0 - c_{\text{OM}}} \right). \tag{6.5}$$

Several observations can be made based on this result[1]:

1. The optimal durability of an asset increases as the expected revenue per day, u_0, increases. In other words, the more revenues a customer anticipates generating from an asset, the longer he or she will want said asset to remain operational. This of course is an intuitive result; Eq. (6.5) provides a quantitative basis for it.

2. The optimal durability decreases with increasing marginal cost of durability of the system. In other words, the more it costs to design a system for an extended period of time, the less it is profitable to do so. Conversely, the cheaper it is to design a system for an extended life ($\alpha << 0$), the more appropriate from a financial standpoint it is to do so.

---

[1] Which exists and is meaningful when $C_0 \alpha < u_0 - c_{\text{OM}}$ and $u_0 > c_{\text{OM}}$.

3. The optimal durability decreases as the discount rate increases. In other words, the riskier the investment, the shorter the system's durability should be.

4. The optimal durability decreases with increasing costs to operate and maintain the system. As can be seen from Eq. (6.5), c_{OM} acts upon the optimal durability in the opposite way to the revenues per day u_0. However, under realistic assumptions, the optimal durability will not be sensitive to c_{OM}, as will be shown in the sensitivity analysis subsection.

Before these observations are probed further, the following subsection offers a concrete numerical example to illustrate both the application and the design consequences of Eq. (6.5).

6.3.1 Numerical Example: Communication Satellite

Communication satellites in geostationary orbits on average cost around $240 million to reach orbit – this cost typically breaks down between 40% for the purchasing cost of the satellite, 40% for launch cost, and 20% for insurance cost. What is the optimal design lifetime of the satellite, under the following assumptions?

- The satellite is expected to generate $80,000 per day from the lease of its on-orbit capacity (transponders) to end-users (e.g., cable companies, global corporations, government agencies).
- The cost to operate the satellite typically varies between 5% and 10% of the purchasing cost of the satellite.
- A discount rate of 10% is recommended.
- Engineering and cost analyses indicate that the cost of the satellite has a weak dependence on its durability; a 4% average marginal cost of durability is suggested.

Given these assumptions, Eq. (6.5) indicates that the optimal durability that maximizes the spacecraft NPV is approximately 13 years (note that today, communications satellites are launched with an average design lifetime of 15 years).

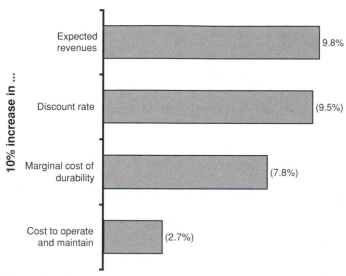

Figure 6.1. Sensitivity analysis of the optimal durability (under nominal conditions from Section 4.2).

6.3.2 Sensitivity Analysis for the Satellite Example

In this subsection, the assumptions underlying the analyses discussed previously are perturbed and the impact on the satellite optimal durability is explored. Four models or parameters affect the solution of Eq. (6.5), namely the system's expected revenue model u_0, its marginal cost of durability $C(T_{\text{Life}})$, the discount rate r, and the cost per year to operate and maintain the system c_{OM}.

The nominal case is the following:

u_0 =   \$80,000/day,

r =   10%,

c_{OM} =   7.5% of purchasing cost of the satellite,

C_0 =   \$150 million,

α =   average marginal cost of durability of 4% increase per year.

The results of the sensitivity analysis are displayed in Figure 6.1. The plot reads as follows: a 10% increase in the satellite's expected revenues, for instance, results in a 9.8% increase in the optimal durability. Similarly, a 10%

increase in the cost to operate and maintain the system results in a 2.7% decrease in the spacecraft's optimal design lifetime.

Analytical expressions for each sensitivity function (not just under one set of nominal conditions) can be obtained by calculating the partial derivatives of optimal durability, as given by Eq. (6.5), with respect to each parameter.

These results indicate where potential customers should conduct careful modeling before selecting design lifetimes for their systems and where they can make do with models of limited accuracy. Of prime importance are market analyses and forecasts of the system's expected revenue (or utility), u_0, as well as financial analysis of the investment riskiness, which in turn will determine the project's discount rate, r. Equally important are the engineering and cost estimate analyses of the system's marginal cost of durability, $C(T_{\text{Life}})$. Of lesser importance to the selection of the optimal durability is the technical analysis and estimation of the cost to operate and maintain the system, c_{OM}.

6.3.3 Discussion

Why do we observe an optimal durability? And what implications does it carry for the design of engineering systems, from a customer's perspective?

An optimal durability exists, under the deterministic and steady-state assumptions described in Section 4, because of discounting and the time value of money (i.e., one dollar spent today is worth more than a dollar generated in a year). For example, in the satellite case, a substantial part of the expenses, $C(T_{\text{Life}})$, is paid prior to launch, whereas the revenues are generated at increasingly later periods. At a sufficiently later period, the revenues generated between T_i and T_{i+1}, when discounted, may become equal to the incremental cost needed to extend the system's durability from T_i and T_{i+1}. If this occurs, a maximum for the system's NPV is obtained, and the durability at which this phenomenon occurs is referred to in this chapter as the optimal durability of the system, as discussed in Section 6.3.

What implication does the existence of an optimal durability carry for the design of engineering systems?

Consider the following: Having conducted a marginal cost of durability for a system, one can define a cost-per-day metric for a system as the ratio of the total system cost to its durability requirement:

$$C_{\text{per\_day}} = \frac{C(T_{\text{Life}})}{T_{\text{Life}}(\text{days})}. \tag{6.6}$$

This definition corresponds to uniformly amortizing the cost of a system over its intended design lifetime. The cost-per-day metric allows a better visualization of the *economies of scale*, if any, that result from extended durability, than the marginal cost of durability. How is that? The marginal cost of durability analysis provides an answer to how capital-intensive durability is for a particular system. Economies of scale (with respect to durability), however, are a measure of how fast the system's cost scales compared with the incremental durability provided for the additional cost. The cost-per-day metric is a good proxy for these economies of scale of durability, because it is a normalized measure of cost per unit time. For example, for the satellite considered in Section 4.1, the cost per day is given in Figure 6.2; it costs $69,000 per day for a 10-year satellite and $50,000 per year for a 15-year satellite. There are $19,000 economies of scale per unit time for the longer-lived satellite (the larger payload satellite in Figure 6.2).

Figure 6.2, which is typical for complex engineering systems, shows an asset whose cost per day decreases monotonically with durability. In other words, the asset exhibits economies of scale in the time dimension, and it is always more cost-effective, on a per-day basis, to extend the durability of the design. *In the absence of other metrics*, this behavior of the cost per day may provide a justification for designers to always push the technical boundary and seek an increasingly longer durability (i.e., formally, there is no "optimal" durability that minimizes the cost per day). Figure 6.2 also suggests that a customer may be better off requesting the manufacturer to provide the maximum achievable durability, given the economies of scale achieved through extended durability.

However, the existence of an optimal durability, as shown previously, indicates that this argument is flawed if the objective in acquiring an engineering system is to maximize its net present value, as opposed to minimizing a cost-per-day or cost per service metric (e.g., selecting the smallest

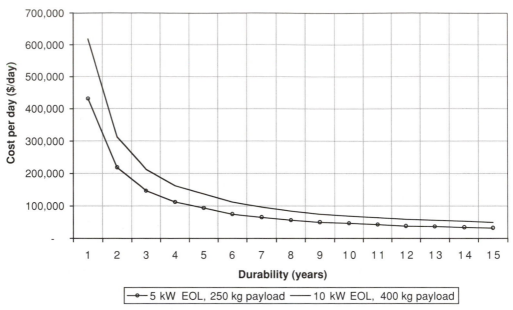

Figure 6.2. Typical communications satellite cost per day.

possible cost per day does not necessarily result in maximizing the system's value). Customers are therefore not necessarily always better off requesting the manufacturer to provide the longest durability technically achievable. Recall that this finding is based solely on the time value of money argument (and on zero marginal cost of durability). This result will be further strengthened when one factors in market/environmental volatility and risk of technology obsolescence, as discussed in the next section.

6.4 Durability, Depreciation, and Obsolescence: A Preliminary Account

The operation research (OR) community has been interested in issues of obsolescence in the context of inventory management and equipment replacement (see for example Song and Zipkin, 1996, David et al., 1997, or Rajagopalan, 1998). This section proposes to bring obsolescence considerations upstream in the design process and account for obsolescence, not in the replacement decision for the equipment, but in the durability selection of the system.

It is intuitively sound for a customer, in selecting a system's durability, to seek to account for the dynamic characteristics of the environment (or market) in which the system is to operate and serve. It feels wrong, for example, to field a system with a 20-year durability when the average life cycles of the technologies embedded in the system are on the order of a couple of years, or when the system is serving a highly volatile market with a risk of dramatically changing in a few years.[2] But how can one formally capture these issues and explore their implications?

Terborgh (1949) articulated an insightful, albeit dramatic, description of the issues hindering a system's actual service life[3]:

> [Machines] live out their mortal span in an atmosphere of combat, a struggle for life. They must defend themselves in a world where species spring up overnight, where the landscape is never twice the same, where the fitful winds of change are never stilled. [In the world of engineered systems] death comes usually by degrees, through a process that may be described as functional degradation. It is a kind of progressive larceny, by which the ever-changing but ever-present competitors of an existing machine rob it of its function, forcing it bit by bit into lower grade and less valuable types of service until there remains at last nothing it can do to justify further existence. (p. 16)

In other words, engineering systems are constantly faced with changes and unpredictability in their environments, as well as functional aggression from competing products. All this tends to *hustle* a system toward obsolescence and shorten its actual service life. System obsolescence is therefore a key concept related to the durability choice problem. The following subsections explore the connection between obsolescence and optimal durability.

6.4.1 Overview of Obsolescence

Obsolescence is an important concept for economists, social scientists, engineers, managers, the medical community, and the operations research

---

[2] The intuition is justified in the (realistic) case of a nonzero marginal cost of durability of the asset.

[3] Can be referred to as durability in an ex post sense, as opposed to the durability (in an ex ante sense) that is discussed in this chapter, namely the intended duration of operation of a system.

community, to name a few. It means and implies different things for different people, and it is important to recognize the multidisciplinary nature of the concept. To discuss any subject matter clearly, it is important to start with a clean and unambiguous set of definitions. It is equally important, however, to keep in mind the objective the particular definition is meant to serve: "Most [. . .] controversies [in the economic literature] about definitions arise from a failure to keep in mind the relation of every definition to the purpose for which it is to be used" (Hicks, 1942).

Obsolescence can be intuitively understood as the loss of value of an artifact because of the development of better ways of providing the same function that the artifact was fulfilling, or because the function itself is no longer desirable or useful. The *Oxford English Dictionary* defines obsolete as "worn out, dilapidated, fallen into disuse," and a second entry defines the term as "persisting but no longer functional or active." I find the following definition more accurate: "an obsolete item is not necessarily broken, worn-out, or otherwise dysfunctional, although these conditions may underscore its obsolescence. Rather the item simply does not measure up to current needs or expectations" (Lemer, 1996).

Obsolescence in an Engineering Context

For engineers, obsolescence revolves primarily around component obsolescence. Its best-known aspect is the nonavailability of parts as suppliers move on to newer products or technology (Solomon et al., 2000). Part obsolescence may occur for a number of reasons. For example, it may be that the demand for the parts under consideration has dropped to such low levels that it is no longer cost-effective to produce them (and, in turn, that low demand for the part may have resulted because a better part is now available on the market). In a similar vein, the Department of Defense, for example, defines obsolescence as diminishing manufacturing sources and material shortages (DMSMS). "DMSMS concerns the loss or impending loss of manufacturers or suppliers of critical items and raw material due to discontinuance of production" (Anonymous, 2000).

> Obsolescence occurs when the last known manufacturer or supplier of an item or raw material gives notice that it intends to cease production.

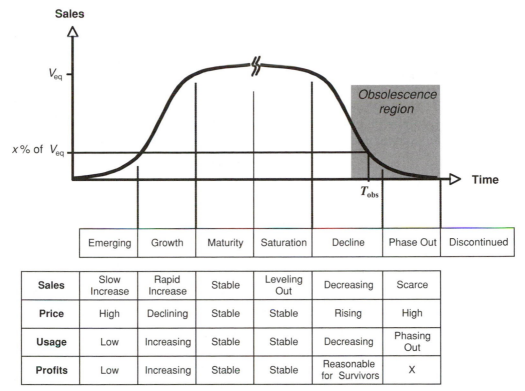

Figure 6.3. Typical component life cycle and phases. Adapted from the ANSI/EIA-724–97 Product Life Cycle Data Model (Hatch, 2000).

The majority of these cases have been in the electronics area, however obsolescence problems affect all weapon systems and material categories. Obsolescence may occur at any phase in the acquisition cycle, from design and development through post-production, and have the potential to severely impact weapon systems supportability and life-cycle costs (Fitzhugh, 1998, p. 170).

Figure 6.3 shows a typical component's lifecycle and progress towards obsolescence, and introduces the concept of *time to obsolescence*.

One way of defining the time to obsolescence of a component, T_{obs}, is in relation to sales volumes, as illustrated in Figure 6.3. Instead of confining obsolescence to the instance when the component is no longer manufactured (or in demand), the time to obsolescence can be defined as that when production (or demand) for a given item drops below a certain level, say 10% of the sales volume during the market maturity phase (Figure 6.3

shows time to obsolescence defined when sales volume drop below x% of the equilibrium volume, V_{eq}.

It is worth noting that several technical consulting companies exist that offer "obsolescence management solutions": they provide, for example, analysis of obsolescence vulnerability and life cycle projections for which components will become obsolete in the near future, what the current component availability is, where procurement problems might exist, or what the replacement options are. Typical times to obsolescence are given in years as deterministic variables, for example, three years for microprocessors, or two years for gate arrays. It would be more appropriate, however, to model time to obsolescence as a random variable with a carefully chosen probability distribution function.

What the engineering and management literatures have not explored, however, is the relation between component obsolescence and system-level obsolescence. This remains a rich field for academic investigation, with significant engineering and managerial implications.

On a side note, to first order, one can conceive of innovation and obsolescence as the opposite faces of the same coin. The "larger [and faster] the improvement in technology, the greater [and sooner] is the chance of scrapping the old [product] before its physical service life has ended" (Carlaw, 2005). The observation of faster innovation leading to faster obsolescence dates back at least to Schumpeter's "creative destruction" (1939).

Economists and Obsolescence

Whereas engineers are interested in component-level obsolescence (micro-obsolescence), economists are mainly focused on system-level obsolescence (macro-obsolescence).[4] For economists, obsolescence is a "decrease in value of an asset because a new asset is more productive, efficient, or suitable for production" (Fraumeni, 1997). In the particular context of information technology, for example, "the availability of superior machines at lower prices is one of the principal reasons that computers lose value as they age" (Whelan, 2002). Lower prices of new vintage assets, however, need not be the only cause of obsolescence of older assets. New technology embedded

[4] There is definitely a need to connect the two levels.

in new vintages often results in better performance or more reliable[5] assets, thus making the services provided by older assets less desirable and therefore less valuable. As stated by Hulten and Wykoff (1996): "When new vintages of capital are introduced into the market, they often contain new *state of the art* technology. . . . The arrival of new better vintages of capital will depress prices of existing old vintages. . . . This decline in value of old vintage is obsolescence."

Obsolescence can depress a system's value, following the introduction of a new asset with state-of-the-art technology, by two mechanisms (or more likely, a combination of both):

1.   The market finds the services provided by the new asset (more) attractive and partly churns away from the services of the old asset to those of the new one. In this scenario, the utilization rate or load factor of the old asset is reduced, thus reducing the overall value of the asset.[6]
2.   Another scenario can unfold as follows: The market finds the new asset's services attractive: however, in order to better compete and prevent churn, the rental price of the services from the old asset are dropped, thus resulting in a decrease in present value of the system.

With this brief overview of obsolescence in an engineering and economic context, we can now turn our attention back to the problem at hand: durability choice in the face of risk of obsolescence.

6.4.2 Durability Choice and System Obsolescence

Assume that a model exists that relates component obsolescence to the system's obsolescence, or, alternatively, assume that obsolescence prediction can be performed at the system level, and that a time scale of obsolescence affects the system's revenues as follows:

$$u(t) = u_0 \times \exp\left[-\left(\frac{t}{T_{\text{obs}}}\right)^2\right]. \tag{6.7}$$

---

[5]   Or any other desirable engineering characteristic of the asset that the market requires or values.
[6]   Recall that the present value of an asset is a forward-looking calculation determined by the future (values of the) services that the asset would deliver.

Other models can be used for the impact of obsolescence on the system's revenue (or flow of service), such as the geometric depreciation pattern used in accounting practice. The conceptual results to be discussed shortly are robust to the model structure. This subsection makes no claim to numerical accuracy, and realism is sacrificed in this first step in order to facilitate understanding of the problem and conceptual results. The main difference between the model of obsolescence (Eq. 6.7) and the model of component life cycle in Figure 6.3 is the absence of emergence and growth phases. This difference may be justified by the fact that, whereas components are mass-produced, the analysis in this chapter is concerned with complex engineering systems, which often reach full operational capability in a time much shorter than their design lifetime: thus the emergence and growth phases are very short compared with their design lifetime. Equation (6.7) can easily be amended to include a short initial period of rise in the revenues per day; this, however, has little effect on the findings.

The optimal durability in this case is given by the solution to Eq. (6.2) in which $u(t)$ is replaced by the expression from Eq. (6.7). Unfortunately, the solution in this case does not admit a closed-form expression, as was the case in Section 6.3 (but it poses no particular problem for a numerical resolution). The optimal durability that maximizes the system's NPV is given by the solution to Eq. (6.8),

$$\left. \frac{\partial V}{\partial T_{\text{Life}}} \right|_{T_{\text{Life}}^*} = \left\{ u_0 \times \exp\left[-\left(\frac{t}{T_{\text{obs}}} \right)^2 \right] - c_{\text{OM}}(t) \right\} \times e^{-rt} - \frac{\partial C\left(T_{\text{Life}}\right)}{\partial T_{\text{Life}}} = 0,$$

(6.8)

and that verifies

$$\left. \frac{\partial^2 V}{\partial T_{\text{Life}}^2} \right|_{T_{\text{Life}}^*} < 0.$$

(6.9)

The optimal durability of a given asset when the risk of obsolescence is taken into account is smaller than the optimal durability when obsolescence is not accounted for.

6.4.3 Numerical Example and Discussion

Consider the satellite example discussed in subsection 6.4.1, and replace the assumption of constant revenues per unit time by Eq. (6.7), thus capturing

Table 6.1. *Summary of model assumptions*

Model input	Assumption
Revenue (or utility) per unit time	$u(t) = u_0 = $ constant
Cost to operate and maintain the system per unit time	$c_{OM}(t) = c_{OM} = $ constant
	$u_0 >> c_{OM}$
Marginal cost of durability	$C(T_{Life}) = C_0(1 + \alpha T_{Life})$, with $\alpha < 1$[a]

[a] This assumption is needed to indicate that the cost of the system under consideration scales up with durability, but it also exhibits economies of scale (with respect to durability and cost). In other words, the incremental cost of the system scales up more slowly than the incremental durability provided for the additional cost.

the impact of obsolescence on the system's revenue stream. The other assumptions summarized in Table 6.1 are maintained. Figure 6.4 illustrates the NPV of the satellite, and its optimal durability, for three different values of time to obsolescence (5, 10, and 15 years).

Figure 6.4 shows that the optimal durability of the satellite, given the model's assumptions, deceases from eight to three and a half years as the expected time to obsolescence decreases from 15 to 5 years. In other words, the sooner customers expect a system to become obsolete, the shorter they should require its durability to be. Alternatively, if depreciation is taken as a proxy for obsolescence, then the faster (larger) the depreciation rate of the asset's category – as provided by the U.S. Department of Commerce's Bureau of Economic Analysis – the shorter the durability of the asset should be.

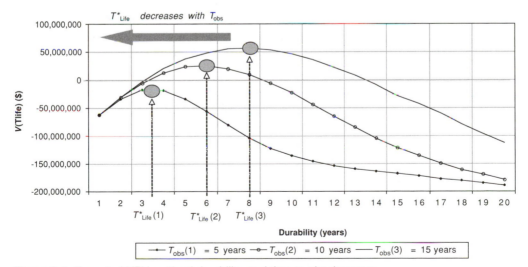

Figure 6.4. Expected NPV, optimal durability, and time to obsolescence.

As an illustration, consider a hypothetical remote sensing satellite with an optical instrument as the payload providing ground images with 15-m resolution. The satellite is designed and launched with a 12-year durability, and its imagery is meant to be commercially available. The satellite's business model assumes strong demand for the satellite's services (i.e., imagery) at the given quality of service (in this case, the optical resolution). The expected revenues outweigh the marginal cost of durability of the spacecraft for 12 years and the cost to operate it for this time period. However, a few years after launch, new, more advanced optical technology becomes available, and a competitor launches a new satellite with a 1-m resolution instrument. The commercial availability of these new images will render the old satellite obsolete and will deal a significant blow to its business model. Furthermore, although the satellite operator paid for a 12-year durability, the operater may have to retire the satellite before the end of its useful life (for lack of demand for its services). Could the operator have made a better choice of durability of the satellite, given the risk of obsolescence? Or is "obsolescence is a risk of essentially the same nature as fire, earthquake, or burglary" that cannot be hedged against (Hotelling, 1925)?[7] Investigating these questions requires the adoption of probabilistic tools and mindset; this is done in the following section.

Caveat: A spacecraft is a special example because, unlike a highway, an aircraft, or a ship, a satellite is, for most cases, not physically accessible, and therefore cannot be maintained, upgraded, or even replenished (therefore, once the propellant is depleted, the spacecraft is out of commission). Hence, there is a stronger mapping for a satellite between the initial design choices and the built-in durability of the asset than with physically accessible and maintainable systems.[8]

6.5 Uncertainty, Risk, and the Durability Choice Problem: A Preliminary Account

Section 6.4 assumed a forecast of a system's utility profile or revenue per unit time, $u(t)$, available as a deterministic variable, and the optimal durability

---

[7] Except perhaps through some form of an insurance policy?
[8] Satellite on-orbit servicing remains a very limited capability confined to low Earth orbit (Saleh et al., 2003).

was computed based on this assumption. In reality, $u(t)$ is related to market volatility, or environmental uncertainties, and cannot be other than a random variable. How should the durability of an asset be selected when its future revenue stream cannot be known with precision but is instead probabilistically distributed?[9] What are the risks, and what is at stake in making a durability choice in the face of uncertainty?

To illustrate the discussion, consider two decision-makers with different risk tolerances faced with a durability choice problem:

1. A risk-averse (RA) decision-maker whose choice of durability is based on some worst-case scenario of the expected revenue stream of the asset, u_{min}. For example, if u is assumed to be normally distributed, with μ and σ for the mean and standard deviation, respectively, then u_{min} can be, for example, equal to $\mu - 2\sigma$ (thus ensuring that the likelihood that the actual revenue stream will fall below u_{min} is very small – 5% under the normal distribution assumption).

2. A risk-taking (RT) decision-maker whose choice of durability is based on some best-case scenario of the expected revenue stream of the asset, u_{max}. For example, if u is assumed to be normally distributed, then u_{max} can be, for example, equal to $\mu + 3\sigma$ (the likelihood that the actual revenue stream will reach u_{max} or above is very small – 5% under the normal distribution assumption).

The durability calculations in Section 6.4 can be conceived of as supporting the durability choice of a risk-neutral decision-maker (working with the average expected revenue stream), and Eq. (6.1) is replaced by Eq. (6.10):

$$E\left[V\left(T_{\text{Life}}\right)\right] = \int_0^{T_{\text{Life}}} \int_0^{\infty} p(u) \times (u - c_{\text{OM}}) \times e^{-rt} du \times dt - C\left(T_{\text{Life}}\right). \quad (6.10)$$

What are the risks faced by each decision-maker in his or her choice of durability?

First, note that the RA decision-maker will choose an asset with a shorter durability than will the RT decision-maker. The shorter-lived asset is optimized for a lower expected utility or revenue stream. Under the

---

[9] Although $u(t)$ more appropriately should be modeled as a stochastic process. This, however, is beyond the focus and scope of this chapter.

assumptions given in Table 6.1, Eq. (6.5) can be used to calculate the difference between the two assets' durabilities:

$$\Delta T = \ln\left(\frac{u_{\max} - c_{\rm OM}}{u_{\min} - c_{\rm OM}}\right). \tag{6.11}$$

The larger the range in the expected utility of the system, the wider the gap between the durability choices of RA and RT.

In addition, the longer-lived asset (of RT), given the marginal cost of durability, will be more expensive than the shorter-lived one (of RA). Under the assumptions given in Table 6.1, the difference in cost between the two assets is given by Eq. (6.12):

$$\Delta C = \frac{C_0 \alpha}{r} \ln\left(\frac{u_{\max} - c_{\rm OM}}{u_{\min} - c_{\rm OM}}\right). \tag{6.12}$$

The RT decision-maker will pay more money up front than RA to acquire the longer-lived asset. Equation (6.12) indicates that the difference in cost increases as the marginal cost of durability of the asset under consideration increases. This incremental investment in durability by RT can be conceived of as a "cost at risk" should events (or the market) unfold unfavorably. In other words, RT would have overinvested in durability, whereas the market could have been better served by a shorter-lived asset.

However, should events (or the market) unfold favorably, the shorter-lived asset (or RA) would fail to capture the incremental present value still available in the market. This consequence of shorter durability will be referred to as a "risk of value forfeited." Under the assumptions given in Table 6.1, the "value forfeited" is given by Eq. (6.13),

$$\Delta PV = \frac{u_{\max} - c_{\rm OM}}{r} \times \left(e^{-r T^*_{u_{\min}}} - e^{-r T^*_{u_{\min}}}\right) = \frac{C_0 \alpha}{r} \times \frac{u_{\max} - u_{\min}}{u_{\min} - c_{\rm OM}}, \tag{6.13}$$

where $T^*_{u_{\min}}$ is the durability optimized given the $u_{\min}$ assumption for the asset's revenue stream.

In summary, there is a dissymmetry in the risks faced by RA and RT in their choices of durability. RA chooses an optimized durability based on a lower expected utility of the asset; in effect, RA chooses a lower durability than does RT. In so doing, RA pays less up front to acquire the asset than RT. Furthermore, the steeper the marginal cost of durability, the more RA saves compared with RT. This is one defining consideration in the choice of durability under uncertainty: the willingness and or ability to pay more (or not) for acquiring a longer-lived asset.

1. On the one hand, should events unfold unfavorably, RT would have incurred a cost penalty for overinvesting in durability, whereas the market could have been better served by a shorter-lived asset. This was referred to previously as the "cost at risk" of the longer-lived asset.

2. On the other hand, should events unfold favorably, RA with the shorter-lived and cheaper asset would be incapable of capturing the remaining present value in the market beyond its durability. In short, by paying less up front for the asset, RA takes the risk of underinvesting in durability and increases the "risk of value forfeited." This is another defining consideration in the choice of durability under uncertainty: the willingness to forfeit incremental present value (that a longer-lived asset would be capable of capturing) and potentially allowing competitors to capture the asset's original market (see Table 2.1 for more details on the implications of reducing or extending a system's durability).

How each scenario, (1) and (2), affects the system's NPV is context- and system-specific. Careful consideration should be given to the various uncertainties and implications before making a cautious or a risky choice of durability. The preliminary account of uncertainty in the choice of durability in this section has avoided describing the NPV implications of different durability choices in order to highlight the two different risks incurred in a cautious versus risky choice of durability.

6.5.1 Numerical Example

Consider the satellite example provided previously, and assume that two discrete expected revenue streams are forecast with equal probability of occurrence, $u_{min} = \$80,000$ per day and $u_{max} = \$120,000$ per day. The risk-averse decision-maker, as defined previously, will make a durability choice based on u_{min}, $T^*_{u_{min}} = 13.5$ years, whereas RT will make a durability choice based on u_{max}, $T^*_{u_{max}} \approx 19$ years. The expected payoffs for each satellite are shown in Table 6.2 (the cost and PV of the shorter-lived satellite are taken as benchmarks).

Table 6.2. *Shorter-versus longer-lived satellites: cost penalties and payoffs*

Scenario/satellite durability	$T^*_{u_{min}} = 13.5$ years	$T^*_{u_{max}} \approx 19$ years
u_{min} unfolds ($p = 50\%$)	C_b	$\Delta C = \$33$ million
	PV_b	$\Delta PV = \$24$ million
		$\Delta NPV >= -\$9$ million
u_{max} unfolds ($p = 50\%$)	C_b	$\Delta C = \$33$ million
	PV'_b	$\Delta PV = \$41$ million
		$\Delta NPV = \$8$ million

Table 6.2 shows that the longer-lived satellite is $33 million more expensive than the shorter-lived one. This incremental cost provides an incremental 5.5 years of life on-orbit. Furthermore, the longer-lived satellite generates either $24 million or $41 million more than the shorter-lived satellite, depending on whether the worst-case (u_{min}) or best-case (u_{max}) market scenario unfolds. This incremental present value translates into a NPV of either –$9 million or $8 million. If the satellite purchasing cost (or any asset under consideration) is considered a sunk cost – as can be objected to the kind of analyses presented in this chapter – then it is obvious that longer-lived assets are always preferred (i.e., the durability choice problem is based on PV, not NPV, considerations).

In the particular case of Table 6.2, the expected NPV calculations show a slight advantage for the shorter-lived satellite (i.e., for the durability choice of the risk-averse decision-maker). Furthermore, should the likelihood of each market scenario be nonsymmetric and skewed toward the worst-case scenario, as would be the case when obsolescence was accounted for, then the case for the shorter-lived asset (and more cautious durability choice) can be made even stronger.

In making the durability choice under uncertainty, numerous considerations should be taken into account (see Section 6.2 and Table 2.1 for a qualitative discussion of some of these issues). This section provides a preliminary account of the financial implications of the durability choice under uncertainty and contends that the financial analysis of durability here developed should be made available to decision-makers to support, in part, the durability choice specification of complex engineering systems.

6.5.2 Limitations

One important limitation in the previous discussion, and in the economic literature on durability, is the assumption that *one choice of durability* has to be made *before the system is designed* and fielded. It is possible, however, to conceive of a design and a scenario by which the potential for multiple durability choices could be embedded in the design of the system. The durability choice can thus be staged over time, and the choice of a specific durability can be postponed until after the system is fielded and some environmental or market uncertainty is resolved. In other words, there would be different points in time when different durability choices could be made (for example, to extend the design lifetime of the system, to upgrade its capabilities, or to modify them). The literature on real options has explored similar concepts in project developments; see for example Trigeorgis (1996) or Amram and Kulatilaka (1999). It would be valuable to promote a similar mindset in the engineering and design community and to investigate the implications of a similar mindset in the durability choice of durable goods from an economic/theoretical perspective (potential for significant practical implications and theoretical findings).

6.6 Conclusions

This chapter addressed the durability choice problem, as seen by the customer, and in the face of network externalities and obsolescence effects. This particular problem has been neglected by economists and engineers, but is poised to become increasingly relevant for program managers and systems engineers, as argued in this chapter.

It is important to distinguish the durability choice problem addressed in this chapter from that discussed in the economics literature. The durability choice problem has received significant attention in the economics literature. Economic studies have focused on the durability choice problem under various market conditions (monopoly or competition), as seen from the manufacturer's perspective. In other words, economists have investigated how manufacturers, under different market structures, choose the durability of their durable goods with the objective of maximizing their profits. It is generally accepted today that monopolists tend to produce goods with inefficient shorter durability than a competitive industry. There are no studies, however, that look at the durability choice from the customer's perspective: if the customer had market power, how would he or she choose the durability of the asset to be acquired? The present chapter differs from the economic literature on durability in that it addresses the durability choice problem from the customer's perspective; that is, an "optimal" durability is sought that maximizes the net present value of a durable asset for the customer (as opposed to maximizing the profits of the manufacturer).

Section 6.2 argued for an augmented perspective on engineering design and optimization as a prerequisite for addressing the durability choice problem of complex engineering systems. More specifically, Section 6.3 made the case that, in order to (quantitatively) discuss issues related to durability, it is important first to conceive of a design, not only as a technical achievement, but also as a value-delivery artifact. And the value (to be) delivered or the flow of service that the system would provide over time, whether tangible or intangible, deserves as much effort to quantify as the system's cost. This perspective was referred to as a value-centric mindset in system design, as opposed to the traditional cost-centric mindset.

Section 6.3 derived analytical results of optimal durability under steady-state and deterministic assumptions. Trends and functional dependence of the optimal durability on various parameters are identified and discussed. A numerical example is provided to illustrate the results and highlight their implications. The existence of an optimal durability challenges the traditional "economies of scale" (in the time dimension) argument in design; for example, selecting the smallest possible cost per day for a system does not necessarily result in maximizing its value. And customers are not necessarily always better off requesting the manufacturer to provide the longest durability technically achievable. This result is further strengthened when market/environmental volatility and risk of technology obsolescence are accounted for, as argued in Section 6.5.

Section 6.4 discussed the durability choice problem under dynamic assumptions, when depreciation and obsolescence are accounted for. First, an overview of the concept of obsolescence is provided in both an engineering and an economic context. Then, the optimal durability of an asset is derived when the risk of obsolescence is taken into account. It is shown, for example, that the sooner customers expect a system to become obsolete (because of a fast-changing underlying technology base, for example), the shorter the optimal durability is. Alternatively, if depreciation is taken as a proxy for obsolescence, then the faster (higher) the depreciation rate of the asset's category, the shorter its optimal durability.

Both Sections 6.3 and 6.4 discussed the durability choice problem under deterministic assumptions. These assumptions are dropped in Section 6.5, and a preliminary account of the durability choice problem under uncertainty is provided in this section. The various risks in making a cautious choice of durability (i.e., short) based on worst-case scenario versus a risky choice of durability (i.e., long) based on optimistic market scenario are identified in Section 6.5. The risks encountered in the durability choice under uncertainty are discussed in Section 6.5 under two headings: the "cost at risk" of the longer-lived asset (should market conditions unfold unfavorably, and for which a shorter-lived asset would have been more appropriate), and the "risk of value forfeited" of the shorter-lived asset (should market conditions unfold favorably).

In summary, this chapter contributes an analytical framework toward a rational choice of durability for engineering systems, as seen from a customer's perspective. It was argued that numerous considerations should be taken into account in the durability choice problem, especially under realistic market and technology evolution conditions. Issues pertaining to the selection of a system's durability are complex and multidisciplinary in nature; they demand careful consideration and require much more attention than they have received so far in the academic literature. This chapter contends that the financial analyses here provided should be made available to decision-makers to support, in part, the durability choice specification of complex engineering systems.

Finally, it should be noted that whether a customer can actually impose his or her choice of durability on the manufacturer or not is an equally important but separate problem from the one investigated in this chapter; it entails discussions of industry structure and market power and possible regulatory considerations. However, whether customers can or cannot impose the choice of durability that maximizes *their* interests, as opposed to that of the manufacturers, should not be an excuse for not investigating what the customer's optimal durability choice should be (this choice can constitute a basis for negotiation with the manufacturer).

────────

This chapter is based on an article written by the author and published in the *Journal of Engineering Design*. Used with permission.

REFERENCES

Amram, M., and Kulatilaka, N. *Real Options: Managing Strategic Investments in an Uncertain World*. Harvard Business School Press, Boston, MA, 1999.

Anonymous. *Program Managers Handbook: Common Practices to Mitigate the Risk of Obsolescence*. Prepared for the Defense Micro-Electronics Activity (DMEA), McClellan Air Force Base, ARINC, 2000.

Carlaw, K. I. "Optimal obsolescence." *Mathematics and Computers in Simulation*, 2005, 69 (1–2), pp. 21–45.

David, I., Greenshtein, E., and Mehrez, A. "A dynamic-programming approach to continuous review obsolescent inventory problems." *Naval Research Logistics*, 1997, 44 (8), pp. 757–74.

Fitzhugh, G. "Rapid retargeting: A solution to electronic systems obsolescence." *IEEE National Aerospace and Electronic Conference, NAECON*, Dayton, OH, 1998, pp. 169–76.

Fraumeni, B. M. "The measurement of depreciation in U.S. national income and product accounts." *Survey of Current Business*, 1997, 77 (7), pp. 7–23.

Gen, M., and Cheng, R. *Genetic Algorithms and Engineering Design*. Wiley, New York, 1997.

Hatch, S. "Diminishing manufacturing sources and material shortages management practices." *Diminishing Manufacturing Sources and Material Shortages Conference*. Jacksonville, FL, 2000, pp. 1–33.

Hicks, J. R. "Maintaining capital intact: A further suggestion." *Economica*, 1942, 9 (34), pp. 174–9.

Hotelling, H. "A general mathematical theory of depreciation." *Journal of the American Statistical Association*, 1925, 20 (151), pp. 340–53.

Hulten, C. H., and Wykoff, F. C. "Issues in the measurement of economic depreciation: Introductory remarks." *Economic Inquiry*, 1996, 34 (1), pp. 10–23.

Lemer, A. C. "Infrastructure obsolescence and design service life." *Journal of Infrastructure Systems*, 1996, 2 (4), pp. 153–61.

Papalambros, P. Y., and Wilde, D. J. *Principles of Optimal Design: Modeling and Computation*, 2nd ed. Cambridge University Press, Cambridge, 2000.

Rajagopalan, S. "Capacity expansion and equipment replacement: A unified approach." *Operations Research*, 1998, 46 (6), pp. 846–57.

Saleh, J. H., Lamassoure, E., Hastings, D., and Newman, D. "Flexibility and the value of on-orbit servicing: A new customer-centric perspective." *Journal of Spacecraft and Rockets*, 2003, 40 (1), pp. 279–91.

Schumpeter, J. A. *Business Cycles: A Theoretical, Historical, and Statistical Analysis of the Capitalist Process*. McGraw–Hill, New York, 1939.

Solomon, R., Sandborn, P. A., and Pecht, M. G. "Electronic part life cycle concepts and obsolescence forecasting." *IEEE Transactions on Components and Packaging Technologies*, 2000, 23 (4), pp. 707–17.

Song, J. S., and Zipkin, H. "Managing inventory with the prospect of obsolescence." *Operations Research*, 1996, 44 (1), pp. 215–22.

Terborgh, G. *Dynamic Equipment Policy*. McGraw–Hill, New York, 1949.

Trigeorgis, L. *Real Options: Managerial Flexibility and Strategy in Resource Allocation*. MIT Press, Cambridge, MA, 1996.

Whelan, K. "Computers, obsolescence, and productivity." *Review of Economics and Statistics*, 2002, 83 (3), pp. 445–61.

Perspectives in Design

The Deacon's Masterpiece and Hundred-Year Aircraft, Spacecraft, and Other Complex Engineering Systems

PREVIEW AND GUIDE TO THE CHAPTER

This text is but a small book about a broad topic. It does not pretend to be exhaustive in its treatment of durability related issues. Important topics such as product replacement or recycling, for example, have not been addressed. More specialized texts would do better justice to these subjects than a summary treatment in the present volume. This Epilogue highlights one particular problem related to durability that has not been addressed, but that is becoming increasingly important.

In 1858, Oliver Wendell Holmes published a poem titled "*The Deacon's Masterpiece or the Wonderful One-Hoss Shay.*" The distinctive feature of the carriage is that all its structural components degrade in such a way that they last a hundred years to a day, and then fail concurrently. Underlying Holmes's poem is a nontrivial design question that is discussed in this paper. To first order, the question can be formulated as follows: How should a system design lifetime be specified, given its underlying components' durability? Or conversely, how should the components in a system be sized given the system's intended duration of operation? A "translation" is undertaken of Holmes's work into engineering parlance, both his sound engineering judgment and his misconception about engineering design. Then, beyond Holmes's example of durability through structural integrity, this paper makes the case for flexibility as an essential attribute for complex engineering designs that can bring about their durability. It is hoped that this Epilogue is read as an invitation to academics and practitioners in engineering disciplines to contribute principles and methodologies for embedding flexibility in the design of complex engineering systems.

1 On Durability through Robustness: The Oliver Wendell Holmes Way

> *Have you heard of the wonderful one-hoss shay,[1]*
> *That was built in such a logical way*
> *It ran a hundred years to a day*
> *And then, of a sudden, it ... ah, but stay,*
> *I'll tell you what happened without delay,*
> *Scaring the parson into fits,*
> *Frightening people out of their wits,*
> *Have you heard of that I say?*

Thus begins Oliver Wendell Holmes's (1809–94) famous poem titled "The Deacon's Masterpiece or the Wonderful One-Hoss Shay," first published in 1858. With a wonderfully engaging style, Holmes takes the reader through the reasoning of a designer, the deacon, in his attempt to build a long-lasting carriage, a *logical design*. But what is the main feature of this design that Holmes characterizes as logical? *Ah, but stay, I'll tell you what happened without delay*! A skillful storyteller, Holmes then takes the reader on a journey that spans a hundred years, accompanying the operation of the one-hoss shay. The deacon, Holmes explains, has built a chaise with no one specific, or localized, weak point: all the structural components of the carriage are equally strong, and degrade in such a way that they last a hundred years to a day, and then fail concurrently, *"all at once, and nothing first, just as bubbles do when they burst."*

Holmes justified the deacon's reasoning as follows: "In a wagon, the weak point is where the axle enters the hub or nave. When the wagon breaks down, three times out of four, I think, it is at this point that the accident occurs. The workman should see to it that this part should never give way; then find the next vulnerable place, and so on, until he arrives logically at the perfect result attained by the deacon." Failure to do so results in systems that break down before wearing out:

> *Now in building of chaise, I tell you what,*
> *There is always somewhere a weakest spot,*
> *In hub, tire, felloe, in spring or thill,*
> *[...]*

---

[1] A colloquial American term for a French *chaise* or carriage pulled by one horse.

In screw, bolt, thorough-brace, lurking still,
Find it somewhere you must and will,
Above or below, within or without,
And that's the reason beyond a doubt,
A chaise breaks down, but doesn't wear out.

Underlying the poem is a serious and nontrivial design issue that transcends the specific case of carriages. To first order, the question can be formulated as follows: how should a system design lifetime be specified, given its underlying components' durability? Or conversely, how should the various components in a system be designed or sized, given the system's intended duration of operation or its design lifetime? Consider the following anecdote: there is an old story, the authenticity of which is contested, that Henry Ford once asked a team of engineers to tour junkyards around the country and examine the conditions of components in discarded Model T Fords. When the team completed the assignment and reported which components had not worn out and had outlasted the useful life of his automobile, Ford is supposed to have instructed his engineers to design these components to lower specification, because they were obviously overbuilt and still had useful life left in them that was wasted when the overall system was discarded; just as the deacon proposed to build his one-hoss shay, each component equally strong (instead of equally weak), so that by the end of the system's operational life, all the components are worn out.

Let us pause for a second and clearly define a few terms of relevance for the discussion of the poem and its engineering implications before proceeding:

Product *durability* and system *design lifetime* are similar in that they both characterize an *artifact's* relationship with *time*. The two expressions, however, sometimes carry different connotations, which, whether warranted or not, deserve nevertheless to be recognized. Product durability has often been used to characterize the durability of a consumer good (a product of limited complexity), whereas design lifetime is often associated with complex engineering systems or capital goods. No such distinction is made in the following pages, and durability and design lifetime are used interchangeably. Furthermore, unless stated otherwise, durability is used in an ex ante sense; that is, it designates the requirement that the manufacturer

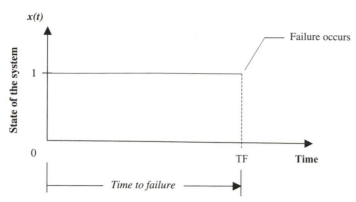

Figure E.1. Relation between the state of a component and its time to failure. Adapted from Rausand and Hyland (2004).

selects for the *intended duration of operation of the system* – this requirement can be imposed on the manufacturer alternatively by the market, the customers, or the regulators. Designers and engineers will make different design choices for the system's structure and components depending on whether the selected design lifetime is, for example, 1 year, 5 years, or 10 years.

2 Time to Failure

The time to failure TF of a component (with one failure mode) is a random variable that describes the time elapsed from when the component becomes operational until it fails for the first time, as shown in Figure E.1.

Neither the design lifetime nor the time to failure need be specified or measured in calendar time; indirect concepts such as operating hours or number of on/off cycles can be substituted instead.

Going back to the deacon, in order to build his carriage *so like in every part*, with no one specific or localized weak point, he proceeded as follows:

[He] inquired of the village folk
Where he could find the strongest oak,
That couldn't be split nor bent nor broke,
That was for spokes and floor and sills;

The other components were equally strong and carefully selected among the toughest of their kind. When the deacon completed his design, he looked at his work with paternal pride, and exclaimed – with a New England pronunciation –

"There!" said the Deacon, "naow she'll dew."

Now that the carriage is fielded and in operation, how does it perform in terms of durability, given its designer's intention for a hundred years' design lifetime? Extraordinary well! But let us have Holmes tell us the story with his characteristic enthusiasm and delight in the deacon's art:

Do! I tell you, I rather guess
She was a wonder, and nothing less!
Colts grew horses, beards turned grey,
Deacon and deaconess dropped away,
Children and grandchildren – where were they?
But there stood the stout old one-hoss shay.

After 100 years of operation, the carriage was still "running as usual, much the same." Unfortunately, on the morning of its 100th year to a day, "there are traces of age in the one-hoss shay, a general flavor of mild decay." But worse was yet to come. Hearing a loud noise, the deacon, perplexed, went to see where the commotion came from; he found

The poor old chaise in a heap or mound,
As if it had been to the mill and ground!
You see, of course, if you're not a dunce,
How it went to pieces all at once,
All at once, and nothing first,
Just as bubbles do when they burst.

End of the wonderful one-hoss shay.
Logic is logic. That's all I say.

And the carriage is no more! The deacon's art is gone, confined to the memory of those who knew the poem.

What the deacon succeeded in doing, from an engineering design point of view, was to predict – with very high accuracy – the time to failure of each component in the carriage. Such an (implicit) analysis and prediction is the realm of the physical approach in reliability, also called structural reliability

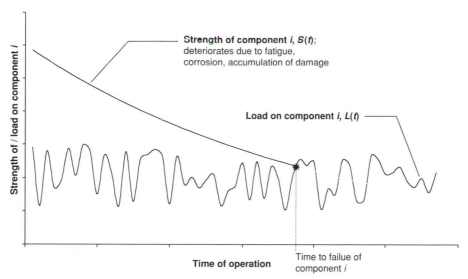

Figure E.2. Time to failure of a component given the load it is supporting, its strength, and its degradation with time. Adapted from Melchers (1999).

analysis. Structural reliability tells us about the ways engineering designs or structures age and fail; their failure depends, on the one hand, on the load supported by each component, and on the other hand, by the material properties and geometric dimensions of each component.

In its simplest expression, structural reliability analysis tells us the following: damage accumulates in a component of a system or a structure as a function of the load it is supporting or the stress levels it is experiencing. The load of course is time-dependent, $L(t)$. The strength of a component, $S(t)$, degrades with time because of a number of failure mechanisms such as fatigue and corrosion. When a critical threshold is exceeded in a given component, the load becoming larger than the strength of the material, the component breaks down. Figure E.2 illustrates one realization of this scenario.

The following expression is the analytical translation of the time to failure of component i as illustrated in Figure E.2 (when the load first exceeds the strength of the material, the component breaks down):

$$\mathrm{TF}_i = \mathrm{Min}\left[t \,|\, S_i(t) < L_i(t)\right]. \tag{1}$$

The strength part, $S_i(t)$: The strength of the component is a function of the material selected to build the component, but it is also a function of the component's cross section and other dimensions, which Holmes and the deacon overlook.

The load part, $L_i(t)$: In addition, in order to accurately predict the time to failure of each component, the deacon would have needed to assess, for its entire intended duration of operation, (1) the weight to be carried by carriage, (2) the road conditions it would travel on, (3) and how the load gets distributed on each component given the carriage's design, or its architecture.

Traditionally, risk associated with time to failure is analyzed by considering the mean time to failure of a component, or its MTTF. And both $S_i(t)$ and $L_i(t)$ are assessed in a probabilistic sense. In other words, the time to failure of a component is a random variable characterized by a probability distribution function, not a deterministic variable, as the poem suggests.

In any case, the deacon succeeded in designing a carriage in which each component failed at the same time; that is (assuming the system is made of n components),

$$\text{TF}_1 = \text{TF}_2 = \text{TF}_3 = \cdots = \text{TF}_n. \tag{2}$$

In this extraordinary special case, it is obvious that the design lifetime of a system is equal to the time to failure of its components; all are identical. However, one should not fail to see that, outside this special case, the design lifetime of a system, T_{Life}, is not necessarily equal to the time to failure of any of its components. Components can be designed and assembled in a way that allows their ease of replacement should they break.

Given this observation, one can define a normalized time vector of the ratios of the time to failure of each component to the overall system design lifetime. Analytically, this can be written as follows:

$$T_{\text{f}} = \left[\frac{\text{TF}_1}{T_{\text{Life}}}; \frac{\text{TF}_2}{T_{\text{Life}}}; \frac{\text{TF}_3}{T_{\text{Life}}}; \ldots; \frac{\text{TF}_n}{T_{\text{Life}}} \right]. \tag{3}$$

This vector captures the tension between the different times scales associated with a design: the time to failure of the components in a system versus the overall system design lifetime.

One can analytically translate the anecdote about Henry Ford discussed above as follows: when Ford supposedly instructed his engineers to design the unfailed components in discarded Model T's to lower specification, he was operating on the premise that no element of the normalized time vector defined above should exceed one:

$$\frac{\mathrm{TF}_i}{T_{\mathrm{Life}}} > 1 \tag{4}$$

is a waste of *component* value (according to the story about Henry Ford)

Conversely, Holmes's poem suggests that any element of our normalized time vector that is smaller than one, i.e., the component failing before the end of the system design lifetime, is a wasteful design:

$$\frac{\mathrm{TF}_i}{T_{\mathrm{Life}}} < 1 \tag{5}$$

is a waste of *system* value (according to Oliver Wendell Holmes).

Perhaps Holmes's main misconception about engineering design (for which he is of course to be forgiven, because he was a Harvard medical professor, and because the elegance of his writing makes up for it!) is the rigid connection he implicitly makes between component failure and system failure: a system fails when any of its components fails, and its design lifetime is dictated by the first failure to occur:

$$\mathrm{Min}\,(\mathrm{TF}_1; \mathrm{TF}_2; \mathrm{TF}_3; \ldots; \mathrm{TF}_n) = T_{\mathrm{Life}} \tag{6}$$

(Holmes's implicit assumption and main misconception about engineering design).

This, of course, need not be the case: a system can be designed in a way that allows the ease of replacement of its failed components, thus in effect decoupling the time to failure of components from the system's design lifetime. Modular designs, for instance, provide this feature. Modular designs have received a fair share of attention in the technical literature; see for example Ulrich and Eppinger (2000), or Baldwin and Clark (1999). Loosely speaking, a modular architecture implies that each function of a product or a system is implemented by one physical element, and the interactions between elements are handled by clear and well-defined interfaces. Modular designs allow ease of change of one physical element in the system without affecting the other elements in the system, and a worn-out component can be replaced without impacting the other components or the overall system. This would have been an alternative design for the deacon's carriage, which would in effect decouple the time to failure of the components from the time to failure of the shay (or its design lifetime):

$$T_{Life} \neq \text{Min} \, (TF_1; TF_2; TF_3; \ldots ; TF_n) \tag{7}$$

in the case of a modular design.

Holmes's misconception about engineering design discussed above is based on a failure to recognize this concept of system architecture (e.g., modularity) and its potential to influence system properties such as reliability and flexibility (sometimes referred to as the "ilities"). However, in Holmes's defense, the poem and its underlying concept of system design are based on the notion of *integral architecture*: "The opposite of a modular architecture is an integral architecture [...] modifications to any one particular component or feature may require extensive redesign of the product" (Ulrich and Eppinger, 2000). The reader is referred to Baldwin and Clark (1999) or Ulrich and Eppinger (2000) for a thorough discussion of product architecture and modular and integral designs.

In summary, *systems architecture can couple or decouple component failure and system time to failure* (or a system's design lifetime). The relationship between a system design lifetime and the time to failure of its underlying components, for integral (coupled) or modular designs (decoupled), is symbolically represented in Figure E.3.

3 Beyond Robustness: On Durability through Flexibility in System Design

Two hundred fifty years ago, the deacon built his carriage, which despite our mild criticism of Holmes's misconception about engineering design remains a "perfectly intelligible conception" with some sound engineering judgment. Much has changed, however, since: five score years after Holmes published his poem, the first Earth orbiting satellite was launched; the integrated circuit (IC) was invented in 1958, paving the way for the rise of computers and the information age; airplanes had been flying since 1903, and, perhaps sadly for the deacon's art, in 1885, Karl Benz designed the first practical automobile powered by an internal combustion engine, and the bell tolled one last time for the one-hoss shay. In short, technological advances had

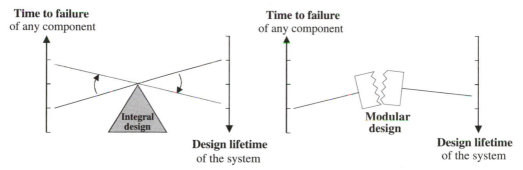

Figure E.3. Illustrative representation of types of relationship between a system's design lifetime and the time to failure of its underlying components for integral and modular designs.

dramatically altered the lives of people since the deacon's art went to pieces, and the complexity of human artifacts had considerably increased since the era of the one-hoss shay. *In this new technological landscape, structural integrity alone no longer guaranteed a product's durability*. Much more is required from a technical system today to last.

George Terborgh, in 1949, articulated an insightful, albeit dramatic, description of the issues hindering products' service life:

> [Machines] live out their mortal span in an atmosphere of combat, a struggle for life. They must defend themselves in a world where species spring up overnight, where the landscape is never twice the same, where the fitful winds of change are never stilled. [In the world of engineered systems] death comes usually by degrees, through a process that may be described as functional degradation. It is a kind of progressive larceny, by which the ever-changing but ever-present competitors of an existing machine rob it of its function, forcing it bit by bit into lower grade and less valuable types of service until there remains at last nothing it can do to justify further existence. (p. 16)

In other words, engineering systems are constantly faced with changes and unpredictability in their environments, as well as functional aggression from competing products. All this tends to *hustle* a system toward obsolescence and shorten its useful operational life (sometimes referred to as *service life*). Systems that thrive longer are the ones that are capable of coping with unpredictability and adapting to changes in their environments.

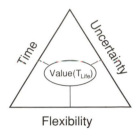

Flexibility

Figure E.4. Triad: time, uncertainty, and flexibility (the T–U–F connections), three faces of the same coin.

I propose that it is precisely this elusive characteristic of a design, its flexibility, that allows a complex engineering system to have an extended operational life and outlive more optimized or rigid designs. Flexibility of a design can be defined as the property of a system that allows it to respond to changes in its initial objectives and requirements occurring after the system has been fielded, in a timely and cost-effective way. Furthermore, I suggest that time, uncertainty, and flexibility are three faces of the same coin in engineering design: on the one hand, time and uncertainty are intrinsically related, for if there were no tomorrow, there would be no uncertainty; and the longer the time horizon considered, the larger the uncertainty. Flexibility, on the other hand, reduces a design's exposure to uncertainty, and thus helps sustain the value of the system over time. As a result, from an analyst's perspective, a flexible design characterizes a system that can cope with uncertainty and changes in its environment; therefore it will have a longer operational lifetime than more rigid or optimized designs. Conversely, from a designer's perspective, if a system is to be fielded with an extended design lifetime, it has to be capable of dealing with significant uncertainty and changes in its environment; therefore flexibility has to be embedded in the system. The nexus of this triad is the value of the system as a function of its design lifetime. The triad consisting of time, uncertainty, and flexibility is graphically represented in Figure E.4.

In summary, it is proposed that the flexibility of a design is a critical property in today's technologically intensive world, the counterpart of structural integrity during the deacon's time, which allows a system to remain usefully operational and outlive rigid (optimized?) designs.

Once flexibility is established as the analog of structural integrity, a good proxy for the functional analog of a part's structural time to failure (TF), discussed in the previous section, is a part's time to obsolescence (TO).

Obsolescence can be intuitively understood as the loss of value of a part because of the development of a better way of providing the same function that the part in question was fulfilling, or because the function itself is no longer desirable or useful. The best-known aspect of technology obsolescence is the nonavailability of parts as suppliers move on to newer technologies and products. Obsolescence impacts all kind of materials and parts, not just electronics, and can occur any time during a system's lifecycle.

Time to obsolescence (TO) of a part or a component can be treated as a random variable with an associated probability distribution function. One measure for this time scale, the time to obsolescence of a part or a component, is when the last known manufacturer or supplier of a part gives notice that it intends to cease production of this part. Another measure of this time scale (TO) can be defined in relation to sales volume: when sales of a part or a component drop below a small percentage of its steady-state sales volume, say below 10% of what the sales volume was during the market maturity phase of the component, the part can be considered to be obsolete. Unlike the time to failure of a component, which is a palpable instance, because the component is physically (or functionally) broken, the time to obsolescence of a part is more elusive and difficult to capture. The two measures described above are at best proxies for this parameter.

For example, Intel introduces major product improvements on the market every 16 to 24 months. The price of the current product is reduced when the new product is introduced; then as the market assimilates the new product, the older product is phased out, typically in 3 to 4 years – this is the product's time to obsolescence. Several companies provide obsolescence analysis and prediction for a variety of parts and components.

Consider now a system made of n components. One can define a normalized time vector, T_{obs}, with the ratio of the time to obsolescence of each component to the system design lifetime. Analytically, this is writen as follows:

$$T_{\text{obs}} = \left[\frac{TO_1}{T_{\text{Life}}}; \frac{TO_2}{T_{\text{Life}}}; \frac{TO_3}{T_{\text{Life}}}; \ldots; \frac{TO_n}{T_{\text{Life}}} \right]. \tag{8}$$

In words, this vector compares how fast each component within a system heads toward obsolescence with the intended duration of operation of the system, T_{obs}, just as the vector T_{f} defined previously with the time to failure of each component captures the tension between the different time scales associated with a design and measures an engineering system's exposure to problems of component obsolescence. How is that? A system is unaffected by obsolescence problems if all its components have their time to obsolescence of the same order as the system's design lifetime:

$$\frac{TO_i}{T_{Life}} \approx 1 \quad \text{for all components } (i = 1 \text{ to } n). \tag{9}$$

This situation, however, would be as extraordinarily special a case as the deacon's masterpiece: all components failing at the same time, or all components becoming obsolete at the same time! More realistically, components within a system will have different times to obsolescence; when any element of our normalized time vector is smaller than one, it flags a problem component that is likely to become obsolete early in the system's operational life:

$$\frac{TO_i}{T_{Life}} < 1. \tag{10}$$

This is the functional analog of the *chaise* breaking down before it is worn out. Engineers or program managers would be well advised to design systems in which "problem components," as identified above, are integrated in the system in a way that allows ease of replacement or upgrade. Preemptive action or design decisions should be taken for such components in order to mitigate the risk of their early obsolescence. This lesson, however, does not seem to have been heard or acted upon properly: today, every aircraft in the U.S. military inventory is reported to have problems of obsolescence and nonavailability of parts or electronic components. The cost of resolving obsolescence problems through redesign is extremely high ($250,000, on average, per circuit redesign according to a study by ARINC, 1999).

Going back to the use of the T_{obs} vector, consider, for example the following: the Boeing B-777 relies on the Intel 80486 microprocessor for its flight management system. It is well known that Intel introduces major product improvements onto the market every 16 to 24 months, and the older product is phased out typically in 3 to 4 years. Assuming the B-777 will remain in service for 30 years, then,

$$\frac{TO_{\mu p}}{T_{Life}} = \frac{3}{30} = 0.1. \tag{11}$$

It is therefore clear that the processor will become obsolete during the aircraft's operational lifetime. Upgrade opportunities therefore will become available that offer improved or new functionality. The aircraft's flight management system (FMS), therefore, should be (or should have been) designed to accommodate changes in its microprocessor in a timely and cost-effective way. In other words, flexibility should be embedded in the design of the FMS.

In summary, there is an increasing tension between present-day complex engineering systems' design lifetimes and the various time scales associated with the obsolescence of their underlying components. On one hand, increased reliability, and budgetary constraints in some cases, has extended systems' design lifetimes. On the other hand, rapid advances in technology have significantly reduced components' time to obsolescence. The result is increasing tension between the different time scales associated with a design. Several obsolescence mitigation strategies have been proposed (after-market sources, emulation, reclaim, etc.), but they address only the symptoms of the problem and are adopted in an ad hoc fashion, after the problem manifests itself.

I suggest instead that deeper architectural solutions with embedded flexibility in the design of complex engineering systems should be sought in order to properly address this rising tension between the different time scales in a design. I also propose that flexibility is an elusive but critical property of a design that allows a complex engineering system to cope with uncertainty and changes in its environment, and thus enjoy an extended operational life and outlive more optimized or rigid designs.

But how does one embed flexibility into the design of complex engineering systems, and what are the tradeoffs associated with designing for flexibility? What is the value of flexibility, and what are the penalties – cost, performance, and risk – if any, associated with it?

4 The New Deacon's Masterpiece: Challenge for Poets and Engineers!

During a recent book-signing event, Margaret Atwood, the author of *The Handmaid's Tale* and *The Blind Assassin*, among others, was asked about the lack of closure in some of her writings. She first answered that only detective stories, she felt, required closure. Then she went on to explain with a wonderful sense of humor that, although she writes for her own pleasure and wants the reader to "participate" in the making of her stories or at least their endings, the real reason for the lack of closure is that often she actually does not know how to end the story.

Her answer, in effect, is quite fitting for the questions raised above on how to embed flexibility into the design of complex engineering systems, and what are the tradeoffs associated with designing for flexibility.

Although there is a growing body of literature on flexibility in a multitude of disciplines, from finance to manufacturing and software design, research in the area of flexibility in system design is still in its infancy. And, unlike the theory of robustness in feedback control systems or the practice of robust design – also known as Tagushi's method – in product development, both disciplines abounding in theoretical and practical results, there is not yet a coherent set of results that demonstrates how to embed flexibility into the design of complex engineering systems, nor how to evaluate it and trade it off against other system attributes such as performance or cost. In fact, this growing field of flexibility in system design would benefit most, before practical results are demonstrated or advocated, from a theoretical framework for thinking about issues of flexibility in design.

It is hoped that, just as Ms. Atwood was inviting her readers to participate in the making of her stories, this paper is read as an invitation to academics in engineering disciplines and practitioners to contribute to the growing field of flexibility in system design, and participate in writing and unfolding of the story of flexibility.

Through physical or functional degradation, or loss of economic value, the hand of time lies heavy on human artifacts, components, products, complex engineering systems, and large structures. In addition to "tree and truth" in Holmes's mind:

> *In fact, there's nothing that keeps its youth*
> *So far as I know, but a tree and truth.*
> *(This is a moral that runs at large;*
> *Take it. You're welcome. No extra charge.)*

The author of the New Deacon's Masterpiece can perhaps add "flexible designs"! Will you, reader, be the author of the next chapter in this story?

———————

This epilogue is based on an article written by the author and published in the *ASME Journal of Mechanical Design*. Used with permission.

REFERENCES

Anonymous, "Resolution Cost Factors for Diminishing Manufacturing Sources and Material Shortages." Prepared for the Defense Micro-Electronics Activity (DMEA), McClellan AFB, ARINC, 1999.

Baldwin, C. Y., and Clark, K. B. *Design Rules: The Power of Modularity*. MIT Press, Cambridge, MA, 1999.

Holmes, O. W. *The Autocrat of the Breakfast-Table*. Signet Classics, The New American Library, New York, 1961.

Melchers, R. E. *Structural Reliability Analysis and Prediction*. 2nd ed. Wiley, Chichester, 1999.

Rausand, M., and Hyland, A. *System Reliability Theory: Models, Statistical Methods, and Applications*, 2nd ed. Wiley–Interscience, Hoboken, NJ, 2004.

Terborgh, G. *Dynamic Equipment Policy*. McGraw-Hill, New York, 1949.

Ulrich, K. T., and Eppinger, S. D. *Product Design and Development*, 2nd ed. Irwin McGraw-Hill, Boston, 2000.

Beyond Cost Models, System Utility or Revenue Models

Example of a Communications Satellite

PREVIEW AND GUIDE TO THE APPENDIX

How does one build a utility or revenue model for an engineering system? The solution may not be a simple, universal, one-size-fits-all model. The investigation and building of a revenue model for a particular system is, however, revealing. In this Appendix, the modeling of a communications satellite is considered as an example in order to avoid the two equally unsatisfying choices of a superficial treatment of the subject matter and an abstract analytical exposition of the subject matter.

This Appendix shows how to build revenue models for communications satellites. The motivation for this work is the proposition that satellites, like any other complex engineering systems, should be designed both for technical merit and also as value-delivery artifacts. And the value delivered, or the flow of service that the system delivers over its design lifetime, whether tangible or intangible, deserves as much effort to quantify as the system's cost.

A.1 Introduction

The previous chapters argued that engineering systems should be conceived of not only as technical achievements but also as value delivery artifacts. And the value delivered, or the flow of service that the system would deliver over its design lifetime, whether tangible or intangible, deserves as much effort to quantify as the system's cost. This is one of the main arguments of this work. The following pages elaborate on this proposition. But how does one build a utility or revenue model for an engineering system? And how does this model turn into a value model for the system?

Because the details of a revenue model are system- and context-specific, a satellite example is considered throughout the chapter for illustrative purposes, to highlight the practical implications of the discussion and avoid abstract exposition of the subject matter.

The remainder of this Appendix is organized as follows. Section A.2 highlights the proliferation of engineering systems cost models in the technical literature and contrasts it with the absence of revenue or utility models for such systems. Section A.2 also makes the case for a value-centric mindset in system design (as opposed to a cost-centric mindset). Section A.3 develops the analytics and the model structure for a communications satellite revenue model. The model structure takes into account the size of the payload, the dynamics of the satellite load factor, the service mix or payload usage breakdown (for a system providing different services), and the evolution of the market price for each service provided by the satellite. The main feature of the revenue model structure here developed is the separation of the system's utilization (endogenous variable) from the market valuation of a unit service provided by the system (exogenous variable). Section A.4 provides some models for the satellite loading dynamics (based on historical data). Section A.5 integrates the loading models developed in Section A.4 with the market price for the service provided and builds the communications satellite revenue models sought after in this Appendix. Section A.6 concludes this Appendix.

A.2 Motivation: Proliferation of Cost Models and Absence of Revenue or Utility Models

Cost models are pervasive throughout many industries. In the aerospace industry, for example, almost every engineer or program manager is familiar with cost models or cost estimate relationships (CERs) and has at some point in his or her career performed cost analyses for a system or subsystem. This situation is the laudable result of the emphasis over the past two decades on financial discipline in the design of complex engineering systems. The technical literature saw a proliferation of "design to cost" methodologies in the 1980s, the intellectual underpinning of which rests on the development of cost models for hardware and software at the component, subsystem, and system levels.

In the space sector, for example, several agencies have over the years developed and refined parametric cost models that relate spacecraft cost or subsystem cost to physical, technical, and performance parameters. For example, a spacecraft's cost depends on its size, mass, complexity, technological readiness, design lifetime, and other characteristics. Popular spacecraft cost models used by NASA, the Air Force, and their respective contractors include the NASA/Air Force Cost Model (NAFCOM) and the Aerospace Corporation's Small Satellite Cost Model (SSCM). Spacecraft cost models are based on historical databases of past satellite programs. The basic assumption of parametric cost modeling is that a system or a subsystem will cost the next time what it has cost the previous times.[1]

Engineering systems cost models not only are popular with industry professionals, but also have made their way into the curricula of a number of universities in the United States. Students are being exposed to and educated about cost models, and an increasing number of graduate theses include cost implications among the design tradeoffs being explored. Many of these students will in turn become industry professionals familiar with cost modeling and carry as part of their educational background a sensitivity to cost implications in their technical responsibilities.

So all is well in the best of all possible worlds? Not exactly. Consider the situation in the aerospace industry, for example. Although cost models are pervasive throughout this industry, revenue models or utility models are seemingly nonexistent. It is unfortunate that few aerospace students and industry professionals are familiar with spacecraft revenue or utility models. Why is this the case, and why should it be an issue?

Figure A.1 illustrates the stark contrast between the proliferation of satellite cost models and the absence of satellite revenue or utility models in various sources. The sources ranged from the most technical peer reviewed

---

[1] Stated this way, one can easily see in such an approach both its advantage, of learning from the past, and its limitation, in taking the past as the sole guide for the future (does not allow for breaking the paradigm).

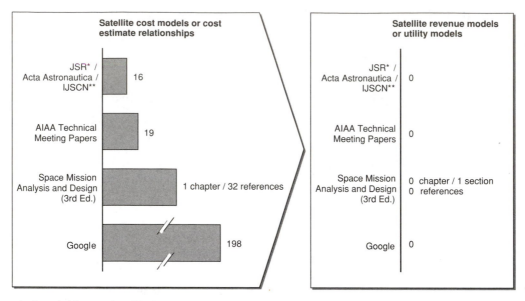

* *Journal of Spacecraft and Rockets*
** *International Journal of Satellite Communications and Networking*

Figure A.1. Proliferation (and lack) of satellite cost and revenue or utility models (as of 02/01/05).

journals to the most popular – unchecked for quality – Google search engine hits. The sources are

- Three leading aerospace journals: the *Journal of Spacecraft and Rockets*, the *International Journal of Satellite Communications and Networking*, *and Acta Astronautica*
- The American Institute for Aeronautics and Astronautics (AIAA) Technical Meeting Papers
- One of the most popular textbooks in the field, *Space Mission Analysis and Design*, edited by Wertz and Larson (1999)
- The Internet, probed by the popular search engine Google

The left-hand side of Figure A.1 constitutes both the cause and consequence of the familiarity of aerospace students and industry professionals with satellite cost models: many cost models are available and discussed in the technical literature. The right-hand side explains why no one is familiar with satellite revenue or utility models: such models simply do not exist. Why should this be an issue?

Although the proliferation of cost models is laudable, the absence of interest in satellite utility or revenue models is troubling. The situation conveys the false impression that satellites are either cost sinks or expensive artifacts whose value or utility profile over their design lifetime is difficult to quantify and does not warrant efforts to do so. More importantly, *the absence of quantitative revenue models* (in the case of commercial systems) *or utility models* (in the case of scientific or military systems) *makes it difficult to build a convincing case for such systems to policy makers or decision-makers*, especially in the light of their exorbitant costs. Furthermore, *the specification and selection of a system design lifetime, or of a system life extension* (e.g., Hubble Space Telescope) *will always have weak arguments fraught with subjectivity in the absence of quantitative revenue/utility models*. The same argument applies for a host of other complex engineering systems.

From an educational point of view, I contend that it is important to provide students with a value-centric perspective on the design of complex engineering systems that incorporates both cost and utility implications of the system over its design lifetime (cost, but also utility, should be part of the design tradeoffs that students explore in their thesis work). As discussed in the previous chapter, the distinction between a value-centric mindset in system design and the traditional cost-centric mindset not only is an academic exercise, but also has direct and practical design implications: different design decisions will ensue if one adopts a cost-centric or a value-centric approach to design.

It should be noted that one reason commercial spacecraft revenue models, for example, do not exist in the open literature may be that such models are simply not publicly available: the information and analyses they contain are understandably proprietary, and satellite operators are not eager to share such data, which can be used to target marketing or sales efforts.

A.3 Developing the Revenue Model Structure for a Communications Satellite

How does one build a utility or revenue model for an engineering system? Where does one start and what model structures should one adopt? Consider the case of a communications satellite in geostationary orbit; its revenue model, $u(t)$, depends on the following:

$$u(t) - u(\text{longitude, service price, number of } Tx, \text{ service mix, technology obsolescence, market volatility}, \ldots). \tag{A.1}$$

The spacecraft's longitude or location in its equatorial orbit provides both an indication of the market size the satellite operator can tap into as well as the competitive intensity over this market (which in turn drives the service price). Prime spacecraft locations have traditionally been over the Americas and Europe, as well as transatlantic longitudes. The size of the payload or the number of transponders on board the spacecraft, as well as the service mix (audio, video, data), is also an important parameter that defines a communications satellite's revenue profile. Finally, the volatility of the market the satellite is intended to serve and the obsolescence of the system's technology base have to be factored in when forecasting a satellite's revenue profile as a function of time, $u(t)$.

In the following, I develop the structure of a revenue model for a FSS satellite (fixed satellite services). DBS satellites (direct broadcast services) are not considered in this Appendix.

For a communications satellite, the revenue model $u(t)$ can be expressed as follows:

$$u(t) = \sum_i (\text{Tx}_{i;t} \times P_i),$$

$$n_{\text{Tx\_active}}(t) = \sum_i (\text{Tx}_{i;t}). \tag{A.2}$$

$\sum_i (\text{Tx}_{i,t})$ is the sum of all transponders on board the spacecraft that are active or leased at time t. P_i is the rent price of each active transponder. In words, Eq. (A.2) simply states that the revenues generated by the system at a particular point in time are equal to the sum of the revenues generated by each transponder active at that time. However, the lease prices of transponders in use at one time need not be homogeneous; hence the use of an indexed price, P_i, per transponder at time t. Indeed, the lease price of a transponder, or, equivalently, the revenues generated by a transponder for the satellite operator varies based on the leased bandwidth and the duration of the contract and on satellite operators offering discounts on transponder price based on duration of lease and capacity.

An argument similar to the one discussed above can be used to begin developing a revenue or utility model for other complex engineering systems (starting with the individual unit of service that the system provides).

The average lease price of a transponder on board a spacecraft at time t can be defined as follows:

$$\overline{P}(t) = \frac{\sum_i (\text{Tx}_{i,t} \times P_i)}{n_{\text{Tx\_active}}(t)}.$$ (A.3)

The satellite load factor at one particular point in time is equal to the total number of transponders in use at time t divided by the total number of transponders on board a spacecraft. It is defined as follows:

$$L(t) = \frac{n_{\text{Tx\_active}}(t)}{N_{\text{Tx\_total}}}.$$ (A.4)

$N_{\text{Tx-total}}$ is the total number of transponders on board a spacecraft. Incorporating Eqs. (A.3) and (A.4) into Eq. (A.2) yields the following expression for the expected revenue model of a communications satellite per unit time:

$$u(t) = N_{\text{Tx\_total}} \times L(t) \times \overline{P}(t).$$ (A.5)

Equation (A.5) assumes that the satellite provides only one type of service, and consequently one average price is considered. This of course need not be the case, and Eq. (A.5) can be generalized to account for a variety of services. One can define a parameter $s_i(t)$ as the percent usage of the satellite utilization rate $L(t)$ at any point in time for the specific service s_i. The family of parameters $s_i(t)$ therefore defines the service mix of a satellite or its payload usage breakdown at any point in time (the system can be viewed as providing a portfolio of services s_i). Furthermore, an average lease price per service can be defined, $\overline{P}_{s_i}(t)$. Given these definitions, Eq. (A.5) can be modified to reflect the revenue model structure for a satellite providing different services:

$$u(t) = N_{\text{Tx\_total}} \times \sum_i \left[s_i(t) \times L(t) \times \overline{P}_{s_i}(t) \right]$$
$$= N_{\text{Tx\_total}} \times L(t) \times \sum_i \left[s_i(t) \times \overline{P}_{s_i}(t) \right],$$
$$\sum_i s_i(t) = 1.$$ (A.6)

Figure A.2 provides a tree representation of the model structure in Eq. (A.6) and summarizes the different levers of revenue generation for a communications satellite.

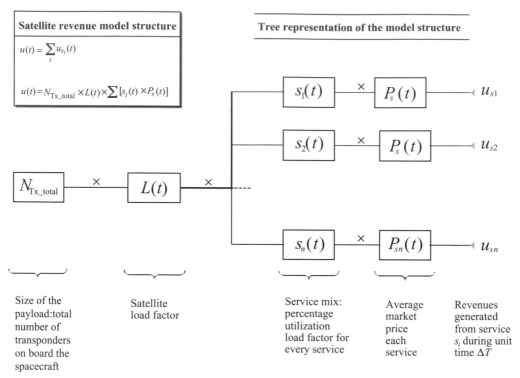

Satellite revenue model structure
$u(t) = \sum_i u_{s_i}(t)$
$u(t) = N_{\text{Tx\_total}} \times L(t) \times \sum [s_i(t) \times P_s(t)]$

Tree representation of the model structure

$N_{\text{Tx\_total}}$ $\times$ $L(t)$ $\times$

$s_1(t)$ $\times$ $P_s(t)$ — u_{s1}

$s_2(t)$ $\times$ $P_s(t)$ — u_{s2}

$s_n(t)$ $\times$ $P_{sn}(t)$ — u_{sn}

Size of the payload:total number of transponders on board the spacecraft

Satellite load factor

Service mix: percentage utilization load factor for every service

Average market price each service

Revenues generated from service s_i during unit time ΔT

Figure A.2. Communication satellite revenue model structure and tree representation.

The model structure provided by Eq. (A.6) highlights four levers of revenue generation: (1) the size of the payload, $N_{\text{Tx-total}}$, (2) the satellite load factor, $L(t)$, (3) the service mix or payload usage breakdown, $s_i(t)$, and (4) the average market price for each service, $\overline{P}_{s_i}(t)$. Each of these levers depends on a variety of parameters. For example, the load factor reflects to some extent how well the satellite operator is managing its on-orbit asset, and how aggressive and successful the operator has been in marketing and selling its on-orbit capacity. In addition to the intrinsic operational effectiveness of the satellite operator, the load factor also reflects the supply/demand characteristics of a given market that the particular satellite is serving. Another lever, the average market price for a service, depends on the one hand on the supply/demand characteristics of a given market for a given service, and on the other hand on the industry structure. Increased overcapacity heightens competitive intensity over a given market and result in increased downward pressure on transponder lease prices. For example, in 2003, the supply/demand imbalance of on-orbit capacity is significantly higher in Central and Eastern Europe (overcapacity of 44%) than in North America (24% overcapacity). This contributed to average lease prices of a transponder in North America in 2003 of $1.2 million/year, whereas in Central and Eastern Europe, transponder lease prices averaged $0.9 million/year. But the supply/demand imbalance is not the only determinant of the average lease price of transponders over a given market; industry structure, or the fragmentation or consolidation of satellite operators in a market, also contributes to the average lease price of transponders. For example, although overcapacity is almost equivalent in North America and Western Europe (roughly 23%), because of the quasi-duopoly of satellite providers in Europe, the lease price of transponders in Western Europe averages $2.1 million/year compared with the $1.2 million/year over the more competitive industry structure in North America in 2003. Table A.1 summarizes the four levers of revenue generation, recapitulates what they depend on, and highlights a few points that satellite operators need to consider before pulling on these levers.

The main feature of the revenue model structure in Eq. (A.6) is the separation of the system's payload utilization (endogenous variable) from the market valuation or price of a unit service provided by the system (exogenous variable). The following section focuses on the endogenous variable in the revenue model provided by Eq. (A.6), namely, the dynamics of the satellite load factor, $L(t)$.

A.4 Modeling Satellite Loading Dynamics

This section focuses on the loading dynamics, or evolution over time of the load factor, of a single satellite after it has been launched, $L(t)$. A satellite load factor can be conceived of as a stochastic process or a random function of time. A stochastic process is simply an indexed family of random variables in which the index corresponds to time.

An argument similar to the one used in building cost models is made in this section, namely looking at historical data, and subsequently identifying and modeling patterns – except that in this section the patterns and regression analysis focus on satellite loading dynamics (instead of cost).

The following is based on a sample of 21 communications satellites, along with their yearly load factor from the time of their launch until their

Table A.1. *Satellite revenue generation levers*

Lever		Dependent upon	Considerations
Size of the payload	$N_{Tx-total}$	• Initial spacecraft requirement and design	• Increasing size of payload may flood the market and result in increased downward pressure on Tx lease price
Satellite load factor	$L(t)$	• Satellite operator sales and marketing efforts • Market supply/demand characteristics	
Service mix or payload usage breakdown (portfolio)	$s_i(t)$	• Satellite operator emphasis on particular services • Market supply/demand characteristics for each service	• Analyze comparative service profitability and volatility • Analyze market unmet demand, competitiveness for each service • Analyze service "stickiness" – difficulty for customers of churning to other providers
Average market price for each service	$P_{s_i}(t)$	• Market structure (e.g., monopoly or competitive market) • Market supply/demand characteristics	• Consolidation increases market power and provides a satellite operator with more control over this lever (market price of service) • Carefully analyze impact of Tx price discount (Will it stimulate demand, i.e., increase load factor? Will it result in a price war with other operators?)

retirement or failure, that were launched between 1980 and 1997 over North America. When the load data in the sample are segmented into three categories defined by the launch period of the satellite, early 1980s, late 1980s, and mid-1990s, and the time axis is initialized to the year of launch, some interesting patterns in the load factor arise. These are discussed in the following subsections.

A.4.1 Load Factors of Satellites Launched in the Late 1980s

Figure A.3 shows the load factor raw data for the first group of satellites launched in the early 1980s. The time axis for all the satellites is initialized to the year of launch.

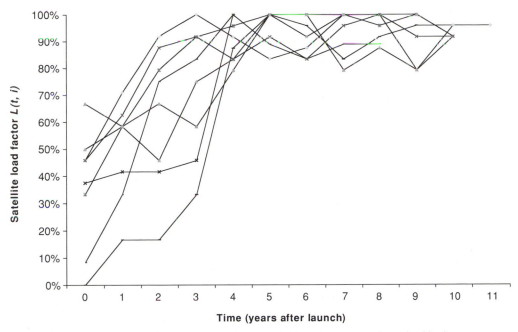

Figure A.3. Load factor raw data for the satellites in the sample that were launched in the early 1980s.

The data show that a satellite's load factor increases after it has been launched, as new customers are acquired and additional transponders are leased. The load factor ramp-up reaches steady state within 3 to 5 years. Interestingly, Figure A.3 shows that some capacity on board a satellite is already prebooked (before the satellite is launched) and the initial average load factor is not 0% (it is in fact 35% for the sample in Figure A.3). This observation makes business sense, and operators ideally should strive to book the entire satellite capacity as soon as or before the spacecraft reaches orbit; failure to do so can be interpreted as an opportunity loss for the operator of the satellite (i.e., an asset or the communications payload in our case is available to generate revenue, but it is not put to work).

The "instantaneous" average load factor is the average at every time step of all the satellite load factors in our sample. It is calculated as follows:

$$L(t) = \frac{1}{n} \sum_{i=1}^{n} L_i(t). \tag{A.7}$$

The subscript i corresponds to satellite i in the database. The instantaneous average load factor of a communications satellite $\overline{L}(t)$ can be modeled using

Table A.2. *Average load factor model parameters for satellites launched in the early 1980s*

Model parameter	Value
Beginning-of-life average load factor, $\overline{L}_{\mathrm{BOL}}$	35%
End-of-life average load factor, $\overline{L}_{\mathrm{EOL}}$	95%
Exponential fill time constant τ	2.5 years
R^2	0.95

three parameters or degrees of freedom: an initial beginning-of-life average load factor $\overline{L}_{\mathrm{BOL}}$ at $t = 0$, a steady-state end-of-life average load factor, $\overline{L}_{\mathrm{EOL}}$, and an exponential fill process with a time constant τ. Equation (A.8) represents the model structure:

$$\overline{L}(t) = \overline{L}_{\mathrm{BOL}} + \left(\overline{L}_{\mathrm{EOL}} - \overline{L}_{\mathrm{BOL}}\right) \times \left(1 - e^{-t/\tau}\right). \tag{A.8}$$

Results of the regression analysis using this model are given in Table A.2.

In addition to the instantaneous average load factor, $\overline{L}(t)$, the data collected allow the modeling of the envelope, or the range within which the satellites load factors fall for every time step after launch. Figure A.4 shows (1) the envelope (minimum and maximum values) of the load factor for the

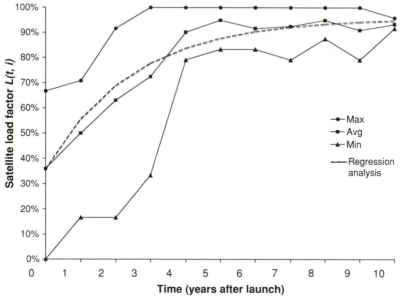

Figure A.4. Load factors (average, min–max, and regression analysis) for the satellites in the sample that were launched in the early 1980s.

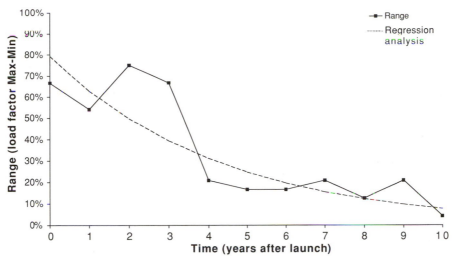

Figure A.5. Dispersion of load factors (Max–Min difference) for satellites launched in the early 1980s.

satellites in our sample, (2) the observed instantaneous average load factor, and (3) the modeled instantaneous average load factor as given by Eq. (A.8) and Table A.2.

Figure A.4 shows an initial large dispersion of load factors right after launch ($\overline{L}_{\text{BOL}}$ varies from 0% to 67%, and has an average of 35%). This may reflect how aggressive a satellite operator has been in prebooking capacity on board its satellite before launch. A satellite with an initial load factor of 0% suggests that the satellite operator has either delayed or not been successful in its sales and marketing effort before its on-orbit asset was launched and became operational. On the other hand, a communications satellite with a high initial load factor suggests either that the operator has been aggressive and successful in its sales efforts prior to the launch of the spacecraft, or that the spacecraft is in fact a "replacement satellite" taking over capacity from another satellite that has reached the end of its service life. This latter hypothesis will be further discussed later.

Figure A.4 also shows that the dispersion of the load factor at every time step narrows down with time and reaches almost a steady state within 4 years. The range, or difference between the minimum and maximum values, in the load factors for satellites launched in the early 1980s is represented in Figure A.5.

Table A.3. *Model parameters for the range of load factors (Eq. (A.9))*

Model parameter	Value
Initial range, r_0	79%
Exponential coefficient, α	0.23
R^2	0.71

This range, $r(t)$, can be modeled as a decreasing exponential function of time. Equation (A.9) represents the model structure (also shown in Figure A.5).

$$r(t) = r_0 \times e^{-\alpha \times t}. \tag{A.9}$$

Results of the regression analysis using this model are given in Table A.3.

As simplification, it is assumed that the range is symmetrical with respect to the sample mean. By doing so, one makes an average error of 18% on the minimum and maximum values of the load factors at each time step for the satellites in the sample (alternatively, one could provide a parametric model for the minimum or maximum values of the load factor).

Alternatively, if the sample and data collected are sufficiently rich, one can propose (and test) various distribution functions for $L(t)$. This unfortunately is not the case in this section and the satellite loading dynamics will be confined to the model proposed by Eqs. (A.8) and (A.9).

A.4.2 Load Factors of Satellites Launched in the Late 1980s

Figure A.6 shows the following for the satellites launched in the late 1980s: (1) the envelope (minimum and maximum values) of the load factor for the satellites in this second group of our sample, (2) the observed instantaneous average load factor, and (3) the modeled instantaneous average load factor as given by Eq. (A.10).

The fundamental difference in the loading dynamics between the satellites that were launched in the early 1980s (previous section) and those launched in the late 1980s is the absence of a ramp-up phase in the latter, as seen by comparing Figure A.4 and Figures A.6. In other words, the sampled satellites that were launched in the late 1980s start with an initially high load factor ($\overline{L}_{\text{BOL}} = 95\%$), and their load factor remains relatively constant

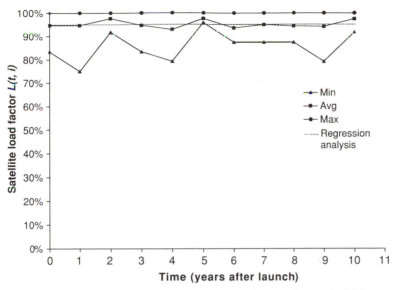

Figure A.6. Load factors (average, min–max, and regression analysis) for seven satellites in our sample that were launched in the late 1980s.

through their design lifetime $(\overline{L}_{BOL} = \overline{L}_{EOL})$, whereas the sampled satellites that were launched in the early 1980s start with a lower load factor $(\overline{L}_{BOL} = 35\%)$ then exhibit a fill process, and take between 3 and 5 years before their load factor reaches a steady state (Figure A.4).

Two reasons can explain this difference in the loading dynamics between these two groups of satellites in the sample: (1) by the late 1980s, satellite operators had determined from their past experience how to aggressively prebook capacity on board their satellites before launch and realized the quantifiable financial advantages of doing so, or (2) most satellites launched in the late 1980s were simply "replacement satellites," taking over capacity from other satellites that were considerably loaded but had reached the end of their service life. If the retiring and replacement satellites have identical capacity, then the beginning-of-life load factor of the replacement satellite will be equal to the end-of-life load factor of the retiring satellite. Otherwise, if the two satellites' capacities differ, one would observe a discontinuity between the L_{EOL} of the retiring satellite and the L_{BOL} of the replacement satellite.

The instantaneous average load factor of the satellites in the sample that were launched in the late 1980s (Figure A.6) can be trivially modeled as a constant. Also, for simplification, it can be assumed that the range or dispersion

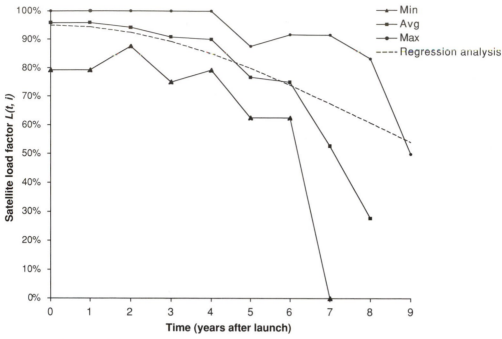

Figure A.7. Load factors (average, min–max, and regression analysis) for satellites in the sample that were launched in the mid-1990s.

of $L(t)$ around the mean $\overline{L}$ is symmetrical with respect to the sample mean. By doing so, one makes an average error of 8% on the minimum values of the load factors at each time step for this second group of satellites in our sample. Mathematically, this trivial model can be written as follows:

$$\overline{L}(t) = \overline{L}_{\mathrm{BOL}} = 95\%,$$
$$r(t) = r_0 = 5\%. \tag{A.10}$$

A.4.3 Load Factors of Satellites Launched in the Mid-1990s

Figure A.7 shows the following for the third category of satellites in the sample (launched in the mid-1990s): (1) the envelope (minimum and maximum values) of the load factor for the satellites in this third group of our sample, (2) the observed instantaneous average load factor, and (3) the modeled instantaneous average load factor, as given by Eq. (A.11).

Figure A.7 shows that satellites in the sample launched in the mid-1990s in the sample start with an initially high load factor ($\overline{L}_{\mathrm{BOL}} = 95\%$), as seen previously in Figure A.6. The same previous interpretation or explanation applies, namely that this reflects either the fact that these satellites are replacement satellites, or that satellite operators are now routinely prebooking most of the capacity on board their satellites before their launch.

Figure A.7 shows, however, one striking difference from all previous load factor dynamics, namely that satellites exhibit a decrease in their load factor after 5 to 7 years of operations. It can be hypothesized (but further analysis is required before this can be confirmed) that this loading pattern corresponds to the onset of obsolescence, when end-users of satellite capacity turn away from aging transponders and switch toward newer, more powerful and reliable units. This hypothesis is plausible given that there has recently been an increasing oversupply of transponders (on-orbit capacity is becoming increasingly commoditized), and end-users have significantly more choice and market power than in the past to "shop around" for newer, better, and cheaper transponders.

The instantaneous average load factor in this case (Figure A.7) can be modeled as a decreasing function of time with two parameters or degrees of freedom: an initial beginning-of-life average load factor $\overline{L}_{\text{BOL}}$, and a time to obsolescence, T_{obs}, as shown in Equation (A.11):

$$\overline{L}(t) = \overline{L}_{\text{BOL}} \times e^{-(t/T_{\text{obs}})^2}. \tag{A.11}$$

Other models of obsolescence can be developed as well, for example:

- One based on the geometric decay pattern used in asset depreciation; see for example Jorgenson (1996) or Hulten and Wykoff (1996).
- One based on the low-pass filter model in signal processing and controls, with at least three degrees of freedom: an initial amplification, a cut-off *time* (which can correspond to the time of obsolescence), and the intensity or slope of the decrease following the onset of obsolescence. In economic parlance, this model would correspond to a concatenation of a one-hoss shay[2] efficiency pattern followed by a decreasing efficiency pattern (straight-line, geometric, or other). It should be noted that this

---

[2] The one-hoss-shay pattern characterizes an asset whose efficiency remains constant throughout its service life, and then fails completely. See the Epilogue for an explanation of the term "shay."

Table A.4. *Model parameters for the average load factor (Eq. (A.11))*

Model parameter	Value
Beginning-of-life average load factor, $\overline{L}_{\mathrm{BOL}}$	95%
Time to obsolescence, T_{obs}	12 years
R^2	0.84

hybrid pattern has not been explored in empirical studies of economic depreciation, but it seems quite promising to address the present limitations with the currently used patterns. This subject, however, falls beyond the scope of this work.

Parameters of the regression analysis using the model in Eq. (A.11) are given in Table A.4.

The range of the data collected for this group of satellites (before the one satellite failure occurred, as seen in Figure A.7) falls within ±15% of the instantaneous average load factor model given in Eq. (A.10) and Table A.4. Unfortunately, the quality of the data for this group of satellites does not warrant further modeling of this range, as was done with the two previous groups of satellites the sample.

A.4.4 Summary of Satellite Loading Dynamics: Four Archetypes

Based on our previous discussion, four archetypes for satellite loading dynamics are proposed. These archetypes are classified based on two dimensions: the type of capacity launched, whether it is a new or replacement satellite; and the market conditions, whether the market is supply-constrained (i.e., the demand can absorb any capacity that is provided) or is competitive and with overcapacity. These four archetypes are represented in Figure A.8.

Archetype A: This archetype or satellite loading pattern corresponds to what is observed with the first group of satellites in the sample (Figures A.3 and A.4), namely an initial ramp-up phase of the load factor followed by a steady-state phase that persists throughout the operational life of the satellite. The satellite load factor increases after launch as new customers are

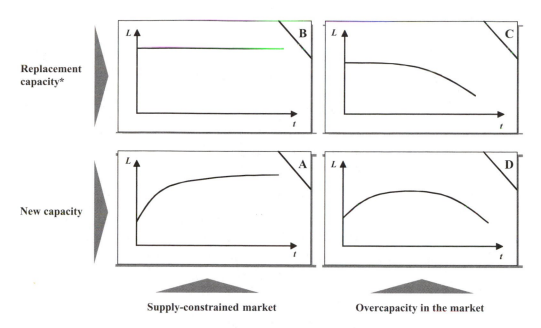

* Or satellite operators with significant experience to prebook most of the capacity on board their satellites before launch

Figure A.8. Satellite loading dynamics: four archetypes classified across two dimensions, type of capacity launched and supply/demand (im)balance in the market.

acquired and additional transponders are leased. The steady-state phase is maintained throughout the operational life of the satellite, as the demand for on-orbit capacity remains unmet (supply-constrained market).

Archetype B: This archetype corresponds to what is observed with the second group of satellites in the sample (Figure A.6), namely a relatively constant load factor throughout the operational life of the satellite (absence of a ramp-up phase). Satellites that exhibit such loading dynamics are replacement satellites taking over capacity from other satellites that are considerably booked but have reached the end of their service life.

Archetype C: This archetype corresponds to what is observed with the third group of satellites in the sample (Figure A.7), namely a steady-state phase with a relatively high beginning-of-life load factor (again with an absence of a ramp-up phase as with archetype B), followed by a decline phase or a decrease in the load after several years of operations. This loading pattern is proposed for replacement satellites that are launched to serve a market that is oversupplied with on-orbit capacity, and where customers

Table A.5. *Worldwide average lease price per year for a 36-MHz transponder*

2000	2001	2002	2003	CAGR 2000–2003
$1.51 million	$1.45 million	$1.37 million	$1.32 million	−4.3%

Data source: Révillion et al. (2004).

can turn away from "aging" transponders and switch toward newer, more powerful and reliable units.

Archetype D: Although this loading pattern was not observed in the data from the sample, its existence can be conjectured for "new" satellites (i.e., not replacement satellites) that are launched to serve a competitive market oversupplied with on-orbit capacity. This archetype therefore has an initial ramp-up phase, a steady-state phase, and a decline phase.

A.5 Integrating Satellite Loading Dynamics with Transponder Lease Price

The previous section investigated the "endogenous" variable in the satellite revenue model provided by Eqs. (A.5) and (A.6), namely the dynamics of the satellite load factor, $L(t)$. This section focuses on the "exogenous" variable in the revenue model, namely the transponder lease price (or price of a unit service provided by the system).

There are different ways for accounting for the average transponder lease price in the revenue model of a satellite, as given by Eq. (A.5) or (A.6). One simple way for doing so is to look at historical data on average transponder lease price in a given market, analyze trends and volatility, and use those data as a proxy for the satellite average transponder lease price, $\overline{P}(t)$, as given in Eq. (A.3). For example, the global average yearly price of a transponder has been steadily declining over the past few years, at approximately 4% per year, because of increased competition and on-orbit overcapacity. The worldwide average lease price of a 36-MHz transponder in 2000 was $1.51 million per year and declined to $1.32 million per year by 2003 (see Table A.5).

For the purposes of the revenue model as given by Eq. (A.5), one can define $\overline{P}(t)$ as a random variable with an appropriately chosen probability distribution function based on price data available, or more elaborately as a

Table A.6. *Twelve revenue models: matrix of loading archetypes with the transponder lease price*

	Transponder lease price evolution over spacecraft design lifetime		
	Worst case	Nominal case	Best case
Loading archetype A	$u_{Aw}(t)$	$u_{An}(t)$	$u_{Ab}(t)$
Loading archetype B	$u_{Bw}(t)$	$u_{Bn}(t)$	$u_{Bb}(t)$
Loading archetype C	$u_{Cw}(t)$	$u_{Cn}(t)$	$u_{Cb}(t)$
Loading archetype D	$u_{Dw}(t)$	$u_{Dn}(t)$	$u_{Db}(t)$

random process (if sufficient data are available to define its autocorrelation function). The following subsection adopts for illustrative purposes a simple model of evolution of $\overline{P}(t)$ based on three scenarios:

- The best-case scenario assumes that the average transponder lease price remains constant throughout the satellite design lifetime.
- The worst-case scenario assumes that the average lease price experiences a steady decrease of -5% per year.
- The nominal case assumes a steady decline of -2.5% per year in average transponder lease price.

Other numbers more carefully chosen for each market can be used instead of the above to define the interval within which $\overline{P}(t)$ is expected to evolve.

A.5.1 The Simple Case: Assuming Lease Price and Load Factor are Independent Variables

In the first simple case, it is assumed that a satellite load factor and its average transponder lease price are independent variables and that the satellite is delivering one type of service. The spacecraft revenue model is readily given by Eq. (A.5).

Twelve revenue models for the matrix of the loading archetypes with the transponder lease price scenario are displayed in Table A.6.

The revenue model $u_{Ab}(t)$, for example, corresponds to the loading dynamics captured by Archetype A and the best-case scenario for the average

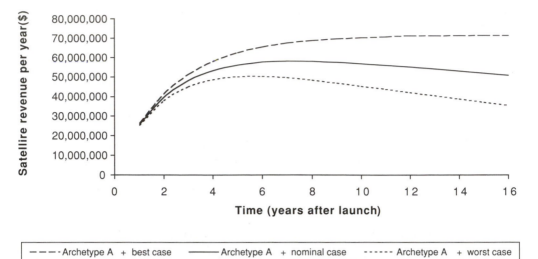

Figure A.9. Three revenue models: $u_{Ab}(t)$, $u_{An}(t)$, $u_{Aw}(t)$). Satellite with fifty 36-MHz transponders and an initial transponder lease price of $1.51 million per year.

transponder lease price. Assuming a satellite with fifty 36-MHz transponders launched in 2000 (average lease price of a transponder $1.51 million/year; the best case scenario assumes this price remains constant throughout the spacecraft design lifetime), $u_{Ab}(t)$ is given by substituting Eq. (A.8) into Eq. (A.5). Numerical results (using the worldwide average lease price from Table A.5) are given by Eq. (A.12), and illustrated in Figure A.9:

$$u_{Ab}(t) = N_{\text{Tx\_total}} \times \overline{L}(t) \times \overline{P}_0,$$
$$L(t) = \overline{L}_{\text{BOL}} + \left(\overline{L}_{\text{EOL}} - \overline{L}_{\text{BOL}}\right) \times \left(1 - e^{-t/\tau}\right),$$
$$u_{Ab}(t) = 50 \times 1.51 \times 10^6 \times \left[0.35 + 0.6 \times \left(1 - e^{-t/2.5}\right)\right]. \qquad \text{(A.12)}$$

The revenue models can also take into account the range or probability distribution of the load factor $\overline{L}(t)$ discussed in Section A.4. One recognizes in Figure A.9, in the best-case scenario, for example, the exponential fill process of the satellite load factor (Eq. (A.8)) multiplied by the average transponder lease price. One notes, for example, the asymptotic behavior of the yearly revenues generated by the satellite as $t \gg \tau$. For the nominal case and worst-case scenario of transponder lease price evolution, note the decline in the revenue generated by the satellite after 5 and 7 years, respectively. The onset of this decline corresponds to the time when the satellite load increase no longer outweighs or compensates for the decline in transponder lease price.

In the more general case, the satellite is delivering multiple services s_i, each with a different price point (and price evolution). The satellite operator, in order to better forecast the revenue profile of its new or replacement on-orbit asset, can assess the potential share of each service and include this information along with the price differential for each service, as given in Eq. (A.6).

It is easy at this point, given the revenue model of the system, or more specifically in the example considered previously, the forecast cash flow that a communications satellite would generate, to calculate the present value of the spacecraft (see Chapter 6). Furthermore, if the marginal cost of durability of the spacecraft is known (as discussed in Chapter 4), this information, along with the revenue model of the system, can be used to develop a value model of the system and make a financially informed decision regarding the system's design lifetime.

A.5.2 Price Discounts and Satellite Load Factor

In the previous subsection, it was assumed that transponder lease price and satellite load factor are independent variables. In addition, the market average lease price of a transponder was used as a proxy for the spacecraft average lease price. The revenue model in this previous simple case was given by the product of the loading dynamics and the average transponder lease price.

In reality, however, it is likely that transponder price discounts (over average market price) spur additional demand and result in an increase of the satellite load factor. Satellite operators do indeed provide discounts to customers based on the duration of the contract and on the leased bandwidth or on-orbit capacity (see for example Restropo and Emiliani, 2004, for an analytic study of this subject). In order to account for the impact of transponder price discounts on the satellite load factor, a metric is introduced and termed the *price elasticity of a satellite load factor*, by analogy with

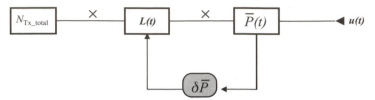

Figure A.10. Representation of a satellite revenue model structure to account for the impact of price changes on the satellite load factor (single service delivered).

the price elasticity of demand or supply in microeconomics. This metric is defined as follows:

$$\varepsilon_{L,P} = \left(\frac{\delta L}{L} \bigg/ \frac{\delta P}{P} \right). \tag{A.13}$$

This metric measures the relative changes in load factor that can be obtained for a given (relative) change in transponder lease price. For example, when $|\varepsilon_{L,P}|$ is large (elastic), the result is interpreted to mean that small changes in transponder lease price significantly impact the satellite load factor. Conversely, when $|\varepsilon_{L,P}|$ is small (inelastic), changes in transponder lease price result in a minor impact on the satellite load factor. It should be noted that $\varepsilon_{L,P}$ is a negative number; a price decrease results in additional demand for on-orbit capacity and therefore an increase in a satellite load factor.

The revenue model structure provided in Eq. (A.5) should then be modified to make explicit this feedback of discounts in transponder lease price on the satellite load factor. Equation (A.14) captures this effect through the intercession of the price elasticity of the load factor metric introduced previously. Figure A.10 provides a modified version of the tree representation of the revenue model structure in the case of a single service delivered to account for this feedback of price changes on the satellite load factor:

$$u(t) = N_{\text{Tx\_total}} \times [L(t) + \delta L] \times \left[\overline{P}(t) + \delta P \right] \Leftrightarrow$$

$$= N_{\text{Tx\_total}} \times \left[L(t) + \varepsilon_{L,P} \frac{\delta P}{P} L(t) \right] \times \left[\overline{P}(t) + \delta P \right] \Leftrightarrow$$

$$= N_{\text{Tx\_total}} \times L(t) \times \overline{P}(t) \times \left[1 + \varepsilon_{L,P} \frac{\delta P}{P} \right] \times \left[1 + \frac{\delta P}{P} \right]. \tag{A.14}$$

Estimating the price elasticity of a satellite load factor is beyond the scope of this work. This metric should be of particular importance to the sales and marketing departments of satellite operators. The metric is market-specific and depends on the supply/demand characteristics as well as the market structure (e.g., monopolistic or competitive market). Satellite operators should be particularly careful in analyzing the impact of transponder price discounts before deciding to pull on this lever, as suggested in Table A.1. In particular, three broad questions should be explored: (1) Will price discounts spur additional demand or is the market oblivious to small price changes? (2) If the market is sensitive to price changes, will the load factor increase compensate for the loss due to price decrease? And (3) if the market is competitive, will price discounts result in a price war with other operators?

A.6 Conclusions

How does one build a utility or revenue model for an engineering system? The question may not admit a universal, one-size-fits-all answer, but a lot can be learned by investigating and building a revenue model for a particular system. This was the premise of this Appendix, and a communications satellite was considered throughout the previous pages for illustrative purposes and in order to avoid two equally unsatisfying choices: a superficial treatment of the subject matter, or an abstract analytical exposition of the subject matter.

This Appendix built revenue models for communications satellites. The motivation for this work is the proposition that satellites, like any other complex engineering systems, should be conceived of not only as technical achievements but also as value-delivery artifacts. And the value delivered, or the flow of services that the system would deliver over its design lifetime, whether tangible or intangible, deserves as much effort to quantify as the system's cost.

This appendix is based on an article written by the author and published in the *International Journal of Satellite Communications and Networking*. Reprinted with permission.

REFERENCES

Hulten, C. H., and Wykoff, F. C. "Issues in the measurement of economic depreciation: Introductory remarks." *Economic Inquiry*, 1996, 34 (1), pp. 10–23.
Jorgenson, D. W. "Empirical studies of depreciation." *Economic Inquiry* 1996, 34 (1), pp. 24–42.

Restropo, J. G., and Emiliani, L. D. "Analysis of current and future states of supply of satellite services in Latin America." *Proceedings of the 23rd AIAA International Communications Satellite Systems Conference. May 9–12, 2004. Monterey, CA.* AIAA 2004–3101.

Révillon, P., Villain, R., Bochinger, S., Gallula, K., Pechberty, M., Rousier, A., and Bellin, S. *World Satellite Communications & Broadcasting Markets Survey, Ten Year Outlook (2004 Edition).* Euroconsult, Paris, August 2004.

Wertz, R., and Larson, W. (eds.). *Space Mission Analysis and Design*, 3rd ed. Microcosm Press, Torrence, CA; Kluwer Academic Publishing, Dordrecht, Boston, London, 1999.

On Durability and Economic Depreciation

PREVIEW AND GUIDE TO THE APPENDIX

Economic depreciation, or the decline in value of an asset with time, has traditionally been associated with the physical deterioration of a capital asset and its functional obsolescence. This Appendix advances two ideas. First, an understanding of depreciation need not be based on deterioration, wear and tear, or obsolescence. Instead, it is shown that the depreciation of a capital asset between two consecutive time periods is equal to the expected incremental present value of the services provided by the asset during this time interval. Second, this Appendix provides a framework for a dynamic forecast of depreciation – a forecast that can be updated and refined as the asset is put to service until its depreciation becomes observable in a resale market. The forecast horizon for depreciation ranges between the expected useful life of an asset and zero. In the latter case, the forecast becomes "observation." Empirical studies of depreciation are a special case within the proposed framework, where the change in an asset's value is "observed" or measured in a resale market.

B.1 Introduction

Depreciation is an important concept for accountants, economists, and financial managers, and it has received significant attention in the literature. The focus in the economics literature has been to a large extent on empirical studies of economic depreciation. These empirical studies investigate aggregate depreciation for cohorts of assets, as opposed to the value trajectory of an individual asset as it ages. This approach has been justified by noting that "most applications in economic growth, production analysis, environmental economics, industry studies and tax analysis are concerned primarily with the average experience of heterogeneous population of capital, and not with the idiosyncratic behavior of an individual asset" (Hulten and Wykoff, 1996).

This Appendix adopts a monadic perspective on depreciation; that is, it focuses on the value trajectory of an individual asset as it ages. In so doing, a number of difficulties encountered by previous studies of depreciation are eliminated. For example, the need to distinguish between the intertemporal understanding of depreciation as the change in value of the same asset at two points in time, and depreciation as understood by the proponents of the productivity school[1] as the difference in value of assets of different vintages at the same point in time, becomes irrelevant. Another difficulty is eliminated by adopting this monadic perspective on depreciation: because only one asset is under consideration, there is no need to account for the retirement pattern of assets in order to correct for bias in the sample, as would happen when a cohort of assets was considered.[2] These simplifications allow us to gain some new insights into depreciation, which will be discussed shortly and which might have not been easily identifiable with the previous difficulties.

This Appendix advances two ideas. First, it proposes a reinterpretation of economic depreciation as the expected incremental present value of an asset between two time periods, while deemphasizing the traditional association of depreciation with deterioration and obsolescence. Second, it provides a framework for a dynamic forecast of depreciation – a forecast that can be updated and refined as an asset is put to service until its depreciation becomes observable in a resale market. The forecast horizon for depreciation ranges between the expected useful life of an asset and zero; in the latter case, the forecast becomes "observation." Empirical studies of depreciation are a special case within the proposed framework, in which the change in an asset's value is "observed" or measured in a resale market.

Before going any further, it is important to distinguish between accounting depreciation and economic depreciation, the latter being the focus of

[1] "Call them the 'productivity school' . . ." Peter Hill (1999), referring to the work on depreciation by Fraumeni (1997), Jorgenson (1996), and Hulten and Wykoff (1996).

[2] "If only observations on surviving assets are included in a sample of used assets prices, estimates of depreciation are biased by excluding observations on assets that have been retired" (Jorgenson, 1996). Hulten and Wykoff seem to have been the first to recognize and account for this problem in Hulten and Wykoff (1981a).

Table B.1. *Accounting versus economic depreciation*

Accounting depreciation	Economic depreciation
Cost-centric calculation	Value-centric calculation
Determined by costs and past expenditure, allocated over the expected useful life of the asset	Forward-looking, determined by the value of future services
Noncash expense; relevant for businesses because it reduces taxable income	Equal to the reduction in present value of an asset between two periods

Two or three estimates are required for the calculation, depending on the allocation method employed:
• Asset's useful life
• Method of allocation to be employed:
  ○ Straight-line method
  ○ Family of accelerated methods
• Asset's salvage value (required for some methods, e.g., the straight-line method)	

this work. Accounting depreciation is a cost-centric calculation involving the allocation or spread of the cost of a capital asset, a past expenditure, over its expected useful life. It is therefore a noncash expense that appears on an income statement and is relevant for accountants because it reduces taxable income. Economic depreciation, on the other hand, is a value-centric calculation: it measures the change in present value of a capital asset as it ages. The present value of an asset, in turn, is a forward-looking calculation determined by the future services that the asset would deliver. The differences between accounting and economic deprecation are summarized in Table B.1.

The concept of economic depreciation originated in the seminal work of Harold Hotelling, published in 1925:

In older treatments of depreciation, the cost, or "theoretical selling price" of the product of a machine was conceived of as determined causally by the addition of a number of items of which depreciation is one.... The simple methods referred to are analogous to the naïve type of economic thought for which the only determiner of price is cost and which fails to consider the equally important role played by demand. The viewpoint of the present treatment is that ... depreciation is defined as the rate of decrease of value [of a property or a machine] ... The first step is, therefore, to set up an expression for the value of the [property or a machine] in terms of value of output, operating costs, and the life of the property. Total depreciation over a period is the difference between the value at the beginning of the period and that at its end. (p. 340)

The rest of this Appendix is organized as follows: Section B.2 reviews the traditional interpretation of economic depreciation and its relation to the physical deterioration of a capital asset and its obsolescence. Section B.3 introduces the model's parameters, assumptions, and notation used in the rest of this Appendix. Section B.4 develops the calculations needed to interpret depreciation in terms of expected incremental present value and proposes a framework for a dynamic forecast of depreciation, taking into account asset deterioration. Section B.5 extends the framework to account for the obsolescence of the asset. Section B.6 contains concluding remarks.

B.2 Depreciation, Deterioration, and Obsolescence: The Traditional Interpretation

Economic depreciation has traditionally been associated with, and sometimes is said to be caused by, the physical deterioration of a capital asset and its functional obsolescence. For example, the U.S. Department of Commerce's Bureau of Economic Analysis defines economic depreciation as "decline in value due to wear and tear, obsolescence, accidental damage, and aging" (Katz and Herman, 1997). Similarly, the System of National Accounts, which consists of an "integrated set of macroeconomic accounts based on internationally agreed concepts, definitions, classifications and accounting rules," defines economic depreciation as "the decline, during the course of the accounting period, in the current value of . . . assets . . . as a result of physical deterioration, normal obsolescence, or normal accidental damage" (System of National Accounts, 1993).[3]

---

[3] To avoid confusion between accounting depreciation and economic depreciation, the latter is called "consumption of fixed capital" in the System of National Accounts. This touches on an important controversy in the theory of capital, the discussion of which is beyond the scope of this work. The interested reader is referred to Triplett (1996) for an exposition of this controversy.

Hulten and Wykoff, authors of seminal empirical studies on depreciation, highlight these two causes of depreciation, physical deterioration, and obsolescence: "Machines wear out, trucks break down, electronic equipment become obsolete. As the physical deterioration and retirement of assets cause the productive capacity to decline, a parallel loss in asset financial value occurs. This depreciation of value is [economic depreciation]" (Hulten and Wykoff, 1996). Jorgenson, on the other hand, emphasizes the physical deterioration of an asset, or, as he refers to it, the decline in its productive capacity or efficiency, as the cause of its depreciation: "[Economic] depreciation reflects both the current decline in efficiency [of a capital good] and the present value of future declines in efficiency" (Jorgenson, 1996). Sometimes, decay is substituted for deterioration: "Capital goods experience decay, which means that the older good can provide fewer services in the current period. In the literature, decay is also called *efficiency decline*" (Triplett, 1996). This physical decay of an asset in turn has valuation implications that result in depreciation. Fraumeni (1997) proposes that depreciation should reflect only the decline in efficiency of an asset as it ages and not include obsolescence. She argues that obsolescence, like pure inflation, should be accounted for in the concept of revaluation of an asset, separate from its depreciation, and makes the case for the "usefulness in separating the effects of obsolescence from those associated with physical deterioration" (Fraumeni, 1997).

In the following sections, it will be shown that an understanding of depreciation need not be based on deterioration, wear and tear, decay, or obsolescence of an asset, as has traditionally been done to date.

B.3 The Model

As mentioned previously, the focus in this Appendix is on the value trajectory of an individual asset as it ages. This leaves us with the intertemporal understanding of economic depreciation, namely that the depreciation of an asset between period i and $i+1$ is equal to the difference in the values of the asset between these two periods. Furthermore, it is assumed in the following that the value of a capital asset is equal to the price a rational investor would be willing to pay for it. This price, in turn, is equal to the expected present value of future services that the asset would provide. For simplicity, present value calculations are substituted for net present value and the need to address operating, and maintenance costs, although important in other applications, are not relevant to the purposes of this work and are thus eliminated. Finally, it is assumed that the asset, whose services are leased or rented (it is not owner-utilized), has a design lifetime of N years; that is, it is designed to remain operational for N years, and then for whatever reason – it is either depleted, too unreliable, unsafe, or uneconomical to operate – it will be retired from operations. In the remainder of this work, expected depreciation will be referred to simply as depreciation unless otherwise stated.

B.3.1 Notation

L_i denotes the load factor or utilization rate of the asset during period i. Of course, the utilization of an asset can and most likely does change with time; hence the time index to L. P_i is the rent price for the services provided by a unit utilization of the asset during period i. For example, consider a communications satellite with a payload that consists of 100 transponders. Each transponder represents 1% of the payload and can be leased for $1.5 million per year. The rent price of a unit utilization of the satellite (P_i) is therefore equal to $150 million per year. Depreciation between period i and $i+1$ is denoted by $D_{i \rightarrow i+1}$. The total present value of the asset from the time it is fielded until its retirement is PV_{total}, and the remaining present value of services after the asset has been in operation for i years is denoted as PV_i. Finally, r is the discount rate. These parameters are collected in Table B.2.

Table B.2. *Summary of the model parameters and notation*

Model parameter	Notation
Load factor or utilization rate of the asset during period i	L_i
Rent price of a unit utilization of the asset during period i	P_i
Depreciation between period i and $i+1$	$D_{i \rightarrow i+1}$
Total present value of the asset from the time it is fielded until its retirement	PV_{total}
Remaining present value of services after the asset has been in operation for i year	PV_i
Discount rate	r

B.3.2 Clarification of the Superscript

As mentioned previously, the depreciation of the asset between period i and $i+1$ is equal to the difference in the values of the asset between these two time periods. Analytically, this can be written as follows:

$$D_{i \to i+1} = \mathrm{PV}_i - \mathrm{PV}_{i+1}. \tag{B.1}$$

But there is a missing piece of information in Eq. (B.1), and in most discussions of depreciation in the literature: the timing of the above calculations – when is this depreciation calculated, and why should this matter?

Consider the following: the price a rational investor would be willing to pay for an asset is equal to the (net) present value of the future services the asset would provide. The notion of future services implies a forecast or estimate. It is therefore important to clarify when the forecast of these future services is conducted. Before a capital asset is purchased, an investor develops an estimate of the revenue profile and present value of the future services that the asset would provide. This estimate can be refined and updated as the asset enters operations. The superscript tp is introduced to denote the time when the estimation is conducted. For example, $\mathrm{PV}_{\mathrm{total}}^{tp=0}$ is the investor's best estimate of the present value of the asset at time zero, that is, before the asset is fielded. Similarly, depreciation, as given in Eq. (B.1), can be calculated or estimated at $tp = 0$, also before the asset is fielded:

$$D_{i \to i+1}^{tp=0} = \mathrm{PV}_i^{tp=0} - \mathrm{PV}_{i+1}^{tp=0}. \tag{B.2}$$

Forecasts of the value of future services are upgraded over time after the asset is fielded. The quality of the forecast improves with time, and so does the estimate of depreciation in future periods. For example, if the present time is j ($j \le i$), one denotes the estimate of depreciation of the asset between period i and $i+1$ as follows:

$$D_{i \to i+1}^{tp=j \le i} = \mathrm{PV}_i^{tp=j \le i} - \mathrm{PV}_{i+1}^{tp=j \le i}. \tag{B.3}$$

Empirical studies of depreciation become a special case of the above notation. The present value of the asset after i years of operation, that is, the value of the forecast of its future services, is observed in a market, and can be written as follows:

$\mathrm{PV}_i^{tp=i}$: *observed* present value of an asset (price in a resale market) at time $tp = i$.

It is worth noting that, even though this present value is measured or observed in a market, it still involves a notion of forecast: this observed present value is the current forecast that rational investors have of the remaining value of the flow of services that the asset would provide.

B.4 Depreciation and Incremental Present Value

B.4.1 Estimation of Depreciation before the Asset Is Fielded (Initial Forecast at $tp = 0$)

The total expected present value of an asset is given by Eq. (B.4):

$$\mathrm{PV}_{\text{total}}^{tp=0} = \sum_{i=1}^{N} \frac{L_i \times P_i}{(1+r)^i}. \tag{B.4}$$

Let us the define the incremental present value of the asset between periods 0 and 1 as follows:

$$\Delta\mathrm{PV}_{0\to1}^{tp=0} = \frac{L_1 \times P_1}{1+r}. \tag{B.5}$$

Equation (B.5) captures the expected present value of the services the asset would provide between periods 0 and 1. More generally, one can define the incremental present value between any two consecutive periods i and $i+1$ as follows:

$$\Delta\mathrm{PV}_{i\to i+1}^{tp=0} = \frac{L_{i+1} \times P_{i+1}}{(1+r)^{i+1}}. \tag{B.6}$$

The total present value of an asset is simply equal to the sum of all the incremental present values of the services the asset provides. Substituting Eqs. (B.6) into Eq. (B.4), one can rewrite the total present value of the asset as follows:

$$\mathrm{PV}_{\text{total}}^{tp=0} = \Delta\mathrm{PV}_{0\to1}^{tp=0} + \Delta\mathrm{PV}_{1\to2}^{tp=0} + \cdots + \Delta\mathrm{PV}_{N-1\to N}^{tp=0} = \sum_{i=1}^{N} \Delta\mathrm{PV}_{i-1\to i}^{tp=0}. \tag{B.7}$$

Figure B.1 represents Eq. (B.7) graphically: it shows how each incremental present value stacks up to add to the total present value of the asset.

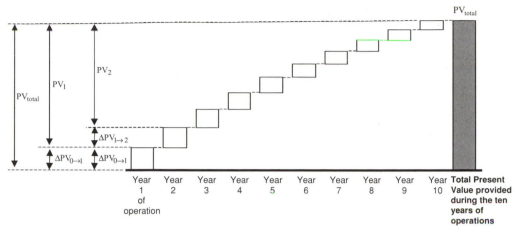

Figure B.1. Total present value and expected incremental present value provided during each period of operation of the asset ($N = 10$ years).

What happens after one year of operation? The remaining expected present value of the asset (still at $tp = 0$) is equal to the value of services provided from periods 1 to period N:

$$PV_1^{tp=0} = \Delta PV_{1\to 2}^{tp=0} + \Delta PV_{2\to 3}^{tp=0} + \cdots + \Delta PV_{N-1\to N}^{tp=0}. \tag{B.8}$$

Using Eqs. (B.7) and (B.8), it is easy to see that the present value of the asset after one year of operation is equal to the total present value of the system minus the incremental present value provided between periods 0 and 1 of operations:

$$PV_1^{tp=0} = PV_{total}^{tp=0} - \Delta PV_{0\to 1}^{tp=0}. \tag{B.9}$$

This relationship (Eq. (B.9)) is illustrated by the set of arrows on the left-hand side of Figure B.1. How is this relevant to depreciation? Depreciation between periods 0 and 1 is equal to the difference in expected present value of the asset when it was first fielded (i.e., total present) and that remaining after one year of operation. This is written as follows:

$$D_{0\to 1}^{tp=0} = PV_{total}^{tp=0} - PV_1^{tp=0}. \tag{B.10}$$

Using Eqs. (B.9) and (B.10), one sees that depreciation between periods 0 and 1 is equal to the incremental present value provided between periods 0 and 1:

$$D_{0\to 1}^{tp=0} = \Delta PV_{0\to 1}^{tp=0}. \tag{B.11}$$

Similarly, the remaining (expected) present value after two years of operations is equal to the sum of the incremental present values from periods 2 to N:

$$
\begin{aligned}
PV_2^{tp=0} &= \Delta PV_{2\to3}^{tp=0} + \Delta PV_{3\to4}^{tp=0} + \cdots + \Delta PV_{N-1\to N}^{tp=0} \\
&= PV_{total}^{tp=0} - \Delta PV_{0\to1}^{tp=0} - \Delta PV_{1\to2}^{tp=0} \\
&= PV_1^{tp=0} - \Delta PV_{1\to2}^{tp=0}.
\end{aligned}
\tag{B.12}
$$

And the depreciation between periods 1 and 2 is

$$
D_{1\to2}^{tp=0} = PV_1^{tp=0} - PV_2^{tp=0} = \Delta PV_{1\to2}^{tp=0}.
\tag{B.13}
$$

More generally, the present value remaining after i periods of operation (again as seen from $tp = 0$) is equal to

$$
\begin{aligned}
PV_i^{tp=0} &= \Delta PV_{i\to i+1}^{tp=0} + \cdots + \Delta PV_{N-1\to N}^{tp=0} \\
&= PV_{total}^{tp=0} - \left(\Delta PV_{0\to1}^{tp=0} + \Delta PV_{1\to2}^{tp=0} + \cdots + \Delta PV_{i-1\to i}^{tp=0} \right) \\
&= PV_{i-1}^{tp=0} - \Delta PV_{i-1\to i}^{tp=0}.
\end{aligned}
\tag{B.14}
$$

And the general expression for (expected) depreciation between period i and $i + 1$ is then given by Eq. (B.15):

$$
D_{i\to i+1}^{tp=0} = PV_i^{tp=0} - PV_{i+1}^{tp=0} = \Delta PV_{i\to i+1}^{tp=0}.
\tag{B.15}
$$

Let us pause for a moment and interpret Eq. (B.15). The result shows that depreciation between two consecutive time periods is equal to the incremental present value of the services provided by the asset during this time interval. To be more precise, it should be noted that, because these calculations are conducted at $tp = 0$, depreciation and present value should be preceded by the qualifier "expected." In any case, Eq. (B.15) shows that there is no need to refer to physical deterioration, wear and tear, or obsolescence of an asset to understand its depreciation. Consider the following thought experiment: assume the existence of an asset that is infinitely robust, that does not deteriorate nor become obsolete with time. This imagined asset would still nevertheless depreciate, and Eq. (B.15) is more capable of explaining its depreciation than is the traditional association of depreciation with deterioration and obsolescence. To summarize, Eq. (B.15) tells us that

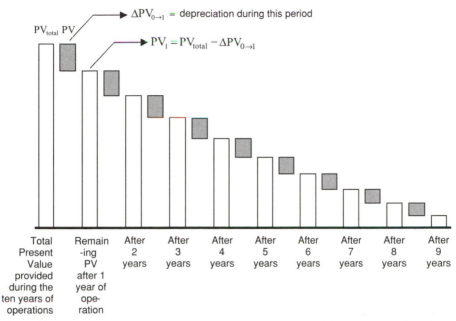

Figure B.2. Incremental present value, depreciation, and remaining present value of an asset after *j* years of operations (*N* = 10 years).

depreciation between two consecutive time periods of operation of a capital asset is equal to the incremental present value of the services provided by the asset during this time interval. This result is illustrated in Figure B.2.

B.4.2 Estimation of Depreciation after the First Period (Forecast at *tp* = 1)

After the first year of operation, the depreciation of the asset during this first time period, $D_{0\to1}^{tp=1}$, can be observed in a market. $D_{0\to1}^{tp=1}$ is therefore an observation of the depreciation of the asset, and it can be compared with the initial estimate of the depreciation, $D_{0\to1}^{tp=0}$, to assess the quality of the initial forecast.

This first year of operation provides new information about the asset, such as its ability to sustain its engineering characteristics, for example, its reliability or performance over time. It also provides new information about the asset's market environment: for example, new information can become available about the demand for the services of the asset, the emergence of competitors with either better or more cost-effective services, raising the specter of obsolescence, and ultimately a better estimate of the price for the asset's services. For simplicity, it is assumed that this information is readily available for all investors to assess the remaining present value of the asset.

Let us introduce now the effect of deterioration. Assume that, after one year of operation, the initial estimate of the market conditions and rental price of a unit utilization of the asset, $\{P_i\}$, remain unchanged. However, assume that the first year of operation revealed that the system has physically deteriorated and could not sustain the initial load, $L_1^{tp=0}$, that it was expected to deliver. Instead, it delivered $L_1^{tp=0} - \delta L_1^{tp=1}$:

$$L_1^{tp=1} = L_1^{tp=0} - \delta L_1^{tp=1}. \tag{B.16}$$

To simplify the notation, the superscript is dropped for the initial estimate of the load, that is, $L_1^{tp=0} \equiv L_1$, and $\delta L_1^{tp=1} \equiv \delta L_1$. This new information can be used to reassess the remaining present value of the asset after the first year of operation by accounting for its physical deterioration. Assuming physical deterioration of the asset δL_i during each time period, its remaining expected present value after the first year of operation can be written as follows:

$$PV_1^{tp=1} = \sum_{i=2}^{N} \frac{(L_i - \delta L_i) \times P_i}{(1+r)^i}. \tag{B.17}$$

If Eq. (B.17) is a rational investor's assessment of the remaining present value of the asset after one year of operation, then the (observed) depreciation of the asset during the first year is equal to

$$D_{0\to1}^{tp=1} = PV_{\text{total}}^{tp=0} - PV_1^{tp=1}. \tag{B.18}$$

Replacing the right-hand side of Eq. (B.18) by Eqs. (B.4) and (B.17) yields

$$
\begin{aligned}
D_{0\to1}^{tp=1} = {} & \left\{ \frac{L_1 \times P_1}{1+r} + \frac{L_2 \times P_2}{(1+r)^2} + \cdots + \frac{L_N \times P_N}{(1+r)^N} \right\} \\
& - \left\{ \frac{(L_2 - \delta L_2) \times P_2}{(1+r)^2} + \frac{(L_3 - \delta L_3) \times P_3}{(1+r)^3} + \cdots + \frac{(L_N - \delta L_N) \times P_N}{(1+r)^N} \right\} \\
= {} & \frac{L_1 \times P_1}{1+r} + \frac{\delta L_2 \times P_2}{(1+r)^2} + \frac{\delta L_3 \times P_3}{(1+r)^3} + \cdots + \frac{\delta L_N \times P_N}{(1+r)^N}.
\end{aligned}
\tag{B.19}
$$

One recognizes in the first term on the right-hand side of Eq. (B.19) the initial assessment of the incremental present value of the asset between periods 0 and 1, as given by Eq. (B.5). Replacing this term (Eq. (B.5)) in Eq. (B.19), one obtains the following:

$$D_{0 \to 1}^{tp=1} = \Delta PV_{0 \to 1}^{tp=0} + \left\{ \frac{\delta L_2 \times P_2}{(1+r)^2} + \frac{\delta L_3 \times P_3}{(1+r)^3} + \cdots + \frac{\delta L_N \times P_N}{(1+r)^N} \right\}. \quad \text{(B.20)}$$

Before the above result is generalized to any two consecutive time periods, let us pause for a moment and interpret it. Equation (B.20) states that the depreciation after the first year of operation is equal to the initial assessment of the incremental present value of the asset between periods 0 and 1 (at $tp = 0$), plus an additional term between braces on the right-hand side of Eq. (B.20) that accounts for the physical deterioration of the asset.

Recall that Eq. (B.11) showed that the initial assessment of the incremental present value of the asset between periods 0 and 1 is also equal to the depreciation at $tp = 0$. Replacing $\Delta PV_{0 \to 1}^{tp=0}$ in Eq. (B.20) by $D_{0 \to 1}^{tp=0}$,

$$D_{0 \to 1}^{tp=1} = D_{0 \to 1}^{tp=0} + \left\{ \frac{\delta L_2 \times P_2}{(1+r)^2} + \frac{\delta L_3 \times P_3}{(1+r)^3} + \cdots + \frac{\delta L_N \times P_N}{(1+r)^N} \right\}. \quad \text{(B.21)}$$

The remaining expected present value of the asset after i periods of operation is the following:

$$PV_i^{tp=1} = \sum_{j=i+1}^{N} \frac{(L_j - \delta L_j) \times P_j}{(1+r)^j}. \quad \text{(B.22)}$$

And the expected depreciation between periods i and $i+1$, estimated at $tp = 1$, is given by

$$D_{i \to i+1}^{tp=1} = PV_i^{tp=1} - PV_{i+1}^{tp=1} = \Delta PV_{i \to i+1}^{tp=1}. \quad \text{(B.23)}$$

The same interpretation of depreciation applies as provided following Eq. (B.15), namely, that the expected depreciation between two consecutive time periods is equal to the incremental present value of the services provided by the asset during this time interval. The difference from the initial assessments (calculations at $tp = 0$) is that the incremental present values at $tp = 1$ are corrected for the physical deterioration of the asset; it is thus easy to show, using Eqs. (B.22) and (B.23), that

$$
\begin{aligned}
D^{tp=1}_{i\to i+1} &= \Delta PV^{tp=1}_{i\to i+1} \\
&= \frac{\left(L^{tp=0}_{i+1} - \delta L^{tp=1}_{i+1}\right) \times P_{i+1}}{(1+r)^{i+1}} \\
&= D^{tp=0}_{i\to i+1} - \frac{\delta L^{tp=1}_{i+1} \times P_{i+1}}{(1+r)^{i+1}}.
\end{aligned}
\tag{B.24}
$$

B.4.3 Estimation of Depreciation after j Periods of Operation (Forecast at $tp = j$)

The results in Eq. (B.24) can now be generalized to any two consecutive time periods. The analytics are identical to the ones proposed in Eqs. (B.22) and (B.23), and the estimated depreciation between i and $i + 1$ calculated at time $tp = j < i$ is given by Eq. (B.25):

$$
\begin{aligned}
D^{tp=j<i}_{i\to i+1} &= \Delta PV^{tp=j}_{i\to i+1} \\
&= \frac{\left(L^{tp=j-1}_{i+1} - \delta L^{tp=j}_{i+1}\right) \times P_{i+1}}{(1+r)^{i+1}} \\
&= D^{tp=j-1}_{i\to i+1} - \frac{\delta L^{tp=j}_{i+1} \times P_{i+1}}{(1+r)^{i+1}}.
\end{aligned}
\tag{B.25}
$$

Equation (B.25) (the third equality) states that the depreciation of an asset between two time periods can be calculated recursively by using the previous estimate of this same depreciation – this is the first term on the second right-hand side of Eq. (B.25) – and updating it by the latest information available about the physical deterioration or performance of the asset – this is the second term on the right-hand side of Eq. (B.25).

Recall that, for all the above calculations, the initial estimate of the rental price of a unit utilization of the asset, $\{P_i\}$, remained unchanged. In other words, it was assumed that the initial estimate ($tp = 0$) of the value of the services was accurate for all subsequent time periods. In Section B.5, this assumption is relaxed and obsolescence effects are introduced.

B.4.4 A Note on Depreciation Patterns, Efficiency Patterns, $\{L_i\}$ and $\{P_i\}$

Empirical studies on depreciation are concerned with collecting market data on asset prices and modeling their depreciation. A summary of such studies up to 1996 is provided in Jorgenson (1996), "starting with the landmark studies of Hulten and Wykoff (1981a, 1981b)."

> Empirical research on depreciation centers on the relationship between the price of an asset and its age, so that assets of different ages or vintages must be analyzed.... Depreciation is the component of unit cost associated with the aging of assets. This component can be isolated by comparing prices of assets of different ages. The objective of empirical research on depreciation is to construct a system of asset prices for this purpose. (p. 24).

An important parameter is introduced in the literature on depreciation; it is sometimes referred to as the relative efficiency or productive capacity of an asset after i years of operation as a fraction of its initial productive capacity,[4] φ_i: The initial efficiency of the asset is normalized to one; that is, $\varphi_1 = 1$ (the initial subscript is set to one rather than zero to maintain consistency with the previous calculations, e.g., in Eq. (B.4)).

It is important to distinguish between efficiency patterns and depreciation patterns. The literature has focused on three types of efficiency patterns for capital assets: the one-hoss-shay, the straight-line, and the geometric pattern.

- The *one-hoss-shay*[5] pattern characterizes an asset whose efficiency remains constant throughout its service life, and then it fails completely. The sequence of its efficiency is given by the following:

$$\varphi_1 = 1; \varphi_2 = 1; \ldots; \varphi_N = 1; \varphi_{N+1} = 0 \text{ (for all } i > N, \varphi_i = 0).$$

---

[4] In some cases, the φ_i are interpreted as "the productive capacity [or efficiency] of an i-year old asset as a fraction of the productive capacity of a newly produced asset" (Hulten and Wykoff, 1981a).

[5] See Epilogue for an explanation of this term and an engineering discussion of Oliver Wendell Holme's poem, "The Deacon's Masterpiece or the Wonderful One-Hoss Shay."

- The *straight-line* pattern characterizes an asset whose efficiency decays linearly with time. The sequence of its efficiency is given by the following:

$$\varphi_1 = 1; \varphi_2 = 1 - 1/N; \varphi_3 = 1 - 2/N; \ldots; \varphi_N = 1 - (N-1)/N; \varphi_{N+1} = 0.$$

- The *geometric* decay pattern characterizes an asset whose efficiency decays geometrically with time. The sequence of its efficiency is given by the following:

$$\varphi_1 = 1; \varphi_2 = 1; -\delta; \varphi_3 = (1-\delta)^2; \ldots; \varphi_N = (1-\delta)^{N-1}$$

with the rate of decay $\delta \in\,]\, 0: 1[$.

These efficiency patterns in turn result in different depreciation patterns. To relate efficiency to depreciation, a model that connects efficiency to present value is needed. Assume the following relation:

$$PV_{total} = \sum_{i=1}^{N} \frac{\varphi_i}{(1+r)^{i-1}} \Delta PV_{0\rightarrow 1}. \tag{B.26}$$

It can be shown using Eqs. (B.15) and (B.26) that the one-hoss-shay efficiency pattern results in the following decline in the present value of the asset:

$$PV_i = PV_0 - \varphi_1 \Delta PV_{0\rightarrow 1} \left\{ 1 + \frac{1}{1+r} + \cdots + \frac{1}{(1+r)^{i-1}} \right\}. \tag{B.27}$$

If the discount rate is set to zero ($r = 0$), Eq. (B.27) shows that the present value declines linearly with time. Hulten and Wykoff (1996) state that "it is well known that the one-hoss shay pattern of efficiency decline implies straight-line depreciation with a zero discount rate." This statement is not accurate because, as is shown in Eq. (B.27), the present value of an asset drops linearly with time (for $r \neq 0$), but depreciation, which is the difference in present value of an asset between two consecutive time periods (see Eq. (B.1)) is constant when the discount rate is zero, that is, it also follows a one-hoss-shay pattern. More generally, the depreciation pattern resulting from a one-hoss-shay efficiency pattern is a concave function given by the following expression (for $r\, 0$):

$$D_{i\rightarrow i+1} = PV_i - PV_{i+1}$$
$$= \frac{\varphi_1}{(1+r)^i} \Delta PV_{0\rightarrow 1}. \tag{B.28}$$

It can be seen from Eq. (B.28) that, for $r = 0$, the depreciation of the asset between any two consecutive time periods is constant:

$$D_{i \to i+1} = \varphi_1 \Delta PV_{0 \to 1} \quad \text{for } 0 \leqslant i < N. \tag{B.28a}$$

Similarly, it can be shown that the depreciation pattern resulting from a straight-line efficiency pattern is the following:

$$D_{i \to i+1} = \frac{\Delta PV_{0 \to 1}}{(1 + r)^i} \left(1 - \frac{i}{N} \right). \tag{B.29}$$

Equation (B.29) also denotes a concave pattern of depreciation (for $r \neq 0$). Setting the discount rate to zero, one notes that the depreciation pattern as given in Eq. (B.29) changes into a straight-line pattern. In other words, a straight-line efficiency pattern results in a straight-line depreciation pattern when the discount rate is set to zero:

$$D_{i \to i+1} = \Delta PV_{0 \to 1} \left(1 - \frac{i}{N} \right) \quad \text{for } r = 0. \tag{B.29a}$$

Finally, it can be shown – and this is a well-known result – that the depreciation pattern resulting from a geometric efficiency pattern is also geometric, and is given by Eq. (B.30):

$$D_{i \to i+1} = \frac{(1 - \delta)^i}{(1 + r)^i} \Delta PV_{0 \to 1}. \tag{B.30}$$

Several empirical studies on depreciation, such as Hulten and Wykoff (1981a, 1981b) and Oliner (1996), have concluded that depreciation patterns for most capital assets are reasonably approximated by the geometric pattern. Katz and Herman (1997), for example, report on an improved methodology for calculating depreciation in the National Income and Product Accounts and state that "the improved methodology uses empirical evidence on prices of used equipment and structures in resale markets, which have shown that depreciation for most types of assets approximate a geometric [as opposed to a straight-line] pattern" (Katz and Herman, 1997). One advantage of the geometric pattern of depreciation is that it requires a single parameter, δ, to characterize the process of economic depreciation. It does not require the assessment of the asset's service life, N, which is notoriously uncertain, especially for long-lived assets, as do the one-hoss-shay and straight-line patterns.

How is the notion of relative efficiency or productive capacity, φ_i, related to the model parameters introduced in this work, namely the load factor or utilization rate of an asset during period i and the rent price of a unit utilization of the asset during period i, L_i and P_i, respectively? If one assumes that the relationship between the relative efficiency and present value of an asset is given by Eq. (B.26), then comparing Eq. (B.4) with Eq. (B.26) (using Eq. (B.5)) provides us with the following equality:

$$\varphi_i = \frac{L_i P_i}{L_1 P_1}. \tag{B.31}$$

Recall that φ_i is defined in the literature as the relative efficiency of an asset after i years of operation as a fraction of its initial productive capacity. Equation (B.31) shows that this parameter is equal to the load factor of an asset times the rent price of its services, divided by the initial load factor times the initial rental price.

B.5 Depreciation and Obsolescence

Obsolescence is an important concept for economists, social scientists, engineers, managers, the medical community, and the operations research community, to name a few. It means and implies different things for different people, and it is important to recognize the multidisciplinary nature of the concept. For economists, obsolescence is a "decrease in value of an asset because a new asset is more productive, efficient, or suitable for production" (Fraumeni, 1997). In the particular context of information technology, for example, "the availability of superior machines at lower prices is one of the principal reasons that computers lose value as they age" (Whelan, 2002). Lower prices of new vintage assets, however, need not be the only cause of obsolescence of older assets. New technology embedded in new vintages often results in better performance or more reliable[6] assets, thus making the services provided by older assets less desirable, and therefore less valuable. As stated by Hulten and Wykoff (1996), "when new vintages of capital are introduced into the market, they often contain new *state of the art*

---

[6]   Or any other desirable engineering characteristic of the asset, which the market requires or values.

technology. . . . The arrival of new better vintages of capital will depress prices of existing old vintages. . . . This decline in value of old vintage is obsolescence."

The question is, by what mechanism(s) does obsolescence impact and depress an asset's value? Based on the model introduced in Section B.3 and Eq. (B.4), which captures the expected present value of an asset, four ways by which obsolescence can decrease an asset's present value are identified:

i.   Decrease in an asset's lifetime: $N_{obs} < N$
ii. Increase in the discount rate: $r_{obs} > r$
iii. Decrease in the asset's utilization or load factor: $L_{i\_obs} < L_i$
iv. Decrease in the rent price of the services provided by the (old) asset: $P_{i\_obs} < P_i$.

The focus in this Appendix will be on the last two mechanisms of obsolescence: the decrease in an asset load factor and the decrease in the rent price of its services. The first two mechanisms, the decrease in an asset lifetime and the increase in the associated investment discount rate, are important topics that require special attention, but that are peripheral to the objectives of this Appendix.

One scenario for the impact of obsolescence on depreciation can unfold as follows: a new asset with state-of-the-art technology is introduced in period i. The market finds its services attractive and partly churns away from the services of the old asset to those of the new one. Therefore, instead of the load factor L_i, the old asset finds itself renting only $L_i - \delta L_{i\_obs}$. Replacing L_i by $L_i - \delta L_{i\_obs}$, in the calculation of an asset's present value (Eq. (B.4)) one can relate the impact of obsolescence on the asset's decrease in present value, and therefore on its depreciation. Another scenario can unfold as follows, relating to the impact of obsolescence on depreciation: again a new asset with state-of-the-art technology is introduced in period i. The market finds its services attractive; however, in order to prevent churn, the rent price of the services from the old asset is dropped from P_i by $P_i - \delta P_{i\_obs}$. Replacing P_i by $P_i - \delta P_{i\_obs}$, in Eq. (B.4) one can relate, as in the previous scenario, the impact of obsolescence on the asset's decrease in present value, and therefore on its depreciation. Given the symmetry of Eq. (B.4) with respect to L and P, "it is immaterial whether the decline in the present value of the remaining services is due to factors which affect the expected quantities of the services or their expected prices. The decline counts as depreciation in both cases" (Hill, 1999).

It is likely that the actual impact of obsolescence would be a combination of the two previous scenarios: a decrease in the old asset load factor coupled with a decrease in its service rent price.[7] For modeling purposes, and in order to avoid confusion with notations used in Section B.4, scenario 2 is adopted and it will be assumed that obsolescence manifests itself by a decrease in service rental price. For example, after one year of operation, the initial estimates of the physical condition of the asset and its load factor, $\{L_i\}$, remain unchanged. However, its service rent price has decreased because of obsolescence:

$$P_1^{tp=1} = P_1^{tp=0} - \delta P_{1\_\text{obs}}^{tp=1}. \tag{B.32}$$

To simplify the notation, the superscript for the initial estimate of the load is dropped; that is, $P_1^{tp=0} \equiv P_1$, and $\delta P_{1\_\text{obs}}^{tp=1} \equiv \delta P_1$. This new information can be used to reassess the remaining present value of the asset after this first year of operation by accounting for its functional obsolescence. Assuming a δP_i price decrease because of foreseen or expected future obsolescence, one can write the remaining expected present value of the asset after the first year of operation as follows:

$$\text{PV}_1^{tp=1} = \sum_{i=2}^{N} \frac{L_i \times (P_i - \delta P_i)}{(1 + r)^i}. \tag{B.33}$$

If Eq. (B.33) is a rational investor's assessment of the remaining present value of the asset (now facing the prospects of obsolescence) after one year of operation, then, using Eq. (B.33), the (observed) depreciation of the asset during this first year is equal to

$$
\begin{aligned}
D_{0\to1}^{tp=1} &= \Delta\text{PV}_{0\to1}^{tp=0} + \left\{ \frac{L_2 \times \delta P_2}{(1+r)^2} + \frac{L_3 \times \delta P_3}{(1+r)^3} + \cdots + \frac{L_N \times \delta P_N}{(1+r)^N} \right\} \\
&= D_{0\to1}^{tp=0} + \left\{ \frac{L_2 \times \delta P_2}{(1+r)^2} + \frac{L_3 \times \delta P_3}{(1+r)^3} + \cdots + \frac{L_N \times \delta P_N}{(1+r)^N} \right\}. \tag{B.34}
\end{aligned}
$$

[7] The operator of the old asset can make an optimal choice during each time period, given the price elasticity for the services, to decrease the service price in order to maximize the revenues (LP).

Equation (B.34) states that the depreciation after the first year of operation is equal to the initial assessment of the incremental present value of the asset between period 0 and 1 (at $tp = 0$), plus an additional term between braces on the right-hand side of Eq. (B.34) that accounts for the impact of obsolescence. The general expression for the assessment of the asset's depreciation between any two consecutive time period i and $i+1$, estimated at $tp = 1$, is given by

$$
\begin{aligned}
D_{i \to i+1}^{tp=1} &= \Delta PV_{i \to i+1}^{tp=1} \\
&= \frac{L_{i+1} \times \left(P_{i+1} - \delta P_{i+1} \right)}{(1+r)^{i+1}} \\
&= D_{i \to i+1}^{tp=0} - \frac{L_{i+1} \times \delta P_{i+1}}{(1+r)^{i+1}}.
\end{aligned} \tag{B.35}
$$

The depreciation between two consecutive time periods is equal to the incremental present value of the services provided by the asset during this time interval. The difference from the initial assessments (calculations at $tp = 0$) is that the incremental present values at $tp = 1$ are corrected for the expected obsolescence of the asset. The third equality in Eq. (B.35) states that depreciation of an asset between two time periods can be calculated recursively by using the previous estimate of this same depreciation – this is the first term on the right-hand side of Eq. (B.25) – and updating it by the latest information available about the impact of obsolescence of the rent price of the asset's services – this is the second term on the right-hand side of Eq. (B.35). Finally, the above results can be generalized for a forecast conducted at any time $tp = j < i$:

$$
\begin{aligned}
D_{i \to i+1}^{tp=j<i} &= \Delta PV_{i \to i+1}^{tp=j} \\
&= \frac{L_{i+1} \times \left(P_{i+1}^{tp=j-1} - \delta P_{i+1}^{tp=j} \right)}{(1+r)^{i+1}} \\
&= D_{i \to i+1}^{tp=j-1} - \frac{L_{i+1} \times \delta P_{i+1}^{tp=j}}{(1+r)^{i+1}}.
\end{aligned} \tag{B.36}
$$

B.6 Concluding Remarks

Economic depreciation, or the decline in value of an asset with time, has traditionally been associated with the physical deterioration of a capital asset and its functional obsolescence. Despite the attention that depreciation has received in the literature, a number of gaps remain in the understanding of this concept, and some factual errors, which go unquestioned, are perpetuated in the literature. For example, the statement by Hulten and Wykoff (1996) that "it is well known that the one-hoss shay pattern of efficiency decline implies straight-line depreciation with a zero discount rate" is not accurate: depreciation, which is the difference in present value of an asset between two consecutive time periods, is constant when the discount rate is zero, that is, it also follows a one-hoss-shay pattern (when the discount rate is not zero, the depreciation pattern resulting from a one-hoss-shay efficiency pattern is a concave function, and its expression is given in Section 4.4).

More generally, this Appendix proposed that an understanding of depreciation need not be based on deterioration, wear and tear, or obsolescence, as has been traditionally done to date. Instead, it showed that the depreciation of a capital asset between two consecutive time periods is equal to the expected incremental present value of the services provided by the asset during this time interval. This understanding of depreciation, loosely speaking as the incremental value provided during the previous period of operation, deemphasizes the traditional association of depreciation with deterioration and obsolescence, and provides some advantages in terms of explaining this concept. Consider, for example, the following thought experiment: assume the existence of an asset that is infinitely robust – that is, it does not deteriorate or become obsolete with time. This imagined asset would still nevertheless depreciate, and the interpretation of depreciation provided in this Appendix is more capable of explaining its depreciation than the traditional association of depreciation with deterioration and obsolescence.

Another idea advanced in this Appendix is that of a dynamic forecast of depreciation – a forecast that can be updated and refined as the asset is put to service until its depreciation becomes observable in a resale market. Consider the following: every year of operation of the asset provides new information about the asset such as its ability to sustain its engineering characteristics, for example, its reliability or performance over time. It also provides new information about the asset's market environment: for example, new information can become available about the demand for the services of this asset, the emergence of competitors with either better or more cost-effective services, thus raising the specter of obsolescence, and ultimately a better estimate of the price for the asset's services. This Appendix showed that depreciation of an asset between two time periods can be calculated recursively by using the previous estimate of this same depreciation and updating it by (1) the latest information available about the physical deterioration or performance of the asset and (2) by the latest information available about the impact of obsolescence of the rent price of the asset's services. The forecast horizon for depreciation ranges between the expected useful life of an asset and zero. In the latter case, the forecast becomes "observation." Empirical studies of depreciation become a special case within the proposed framework in which the change in an asset's value is "observed" or measured in a resale market.

REFERENCES

Fraumeni, B. M. "The measurement of depreciation in U.S. national income and product accounts." *Survey of Current Business*, 1997, 77 (7), pp. 7–23.

Hill, P. *Foreseen Obsolescence and Depreciation*. Organisation for Economic Co-operation and Development, 1999. Unpublished manuscript, available online at http://www.oecd.org/dataoecd/13/25/2550240.pdf (accessed 12/10/05).

Hotelling, H. A. "General mathematical theory of depreciation." *Journal of the American Statistical Association*, 1925, 20, pp. 340–53.

Hulten, C. H., and Wykoff, F. C. "The estimation of economic depreciation using vintage asset price: An application of the Box–Cox power transformation." *Journal of Econometrics*, 1981a, 15, pp. 367–96.

Hulten, C. H., and Wykoff, F. C. "The measurement of economic depreciation," in C. H. Hulten (Ed.), *Depreciation, Inflation, and the Taxation of Income from Capital*. The Urban Institute Press, Washington, D.C., 1981b, pp. 81–125.

Hulten, C. H., and Wykoff, F. C. "Issues in the measurement of economic depreciation: Introductory remarks." *Economic Inquiry*, 1996, 34 (1), pp. 10–23.

Jorgenson, D. W. "Empirical studies of depreciation." *Economic Inquiry*, 1996, 34 (1), pp. 24–42.

Katz, A. J., and Herman, S. W. "Improved estimates of fixed reproducible tangible wealth, 1929–1995." *Survey of Current Business*, 1997, 77 (5), pp. 69–76.

Oliner, S. D. "New evidence on the retirement and depreciation of machine tools." *Economic Inquiry*, 1996, 34 (1), pp. 57–77.

System of National Accounts. *The Production Account*. United Nations Statistics Division, VI, 6.179, 1993. Available online at http://unstats.un.org/unsd/sna1993/toctop.asp (accessed 12/10/05)

Triplett, J. E. "Depreciation in production analysis and in income and wealth accounts: Resolution of an old debate." *Economic Inquiry*, 1996, 34 (1), pp. 93–115.

Whelan, K. "Computers, obsolescence, and productivity." *The Review of Economics and Statistics*, 2002, 83 (3), pp. 445–61.

Index

AMARC [Aerospace Maintenance and
Regeneration Center], 5
Apollo program, 5
Archetypes, loading, 76, 162

Barrier to entry, 21
Battery, spacecraft, 65, 66. *See also*
Depth-of-discharge
Benz, Karl, 136. *See also* Internal
combustion
Blade, stainless-steel, 31
Blades, carbon-steel, 31
Bureau of Economic Analysis,
117
Buy-back, 38. *See also*
Precommitments

Cape Canaveral, 5
Cartel, 14, 15, 30
Chamberlin, Edward, 27, 28
Churn, customer, 115
Coase, Ronald, 35, 49, 50. *See also*
Precommitment; Time
inconsistency
Collusion, 30. *See also* Cartel
Competitive market, 90, 93. *See also*
Supply-constrained market
Competitive price, 36, 37

Complexity, 16
Component life cycle, 92
Component replacement, 135
Concarneau, 1, 2
Conservation of mass. *See* Continuity
equation
Consistency, dynamic, 40
Consistent planning, 50. *See also*
Inconsistency, optimal plans
Consumer goods, 16
Consumer surplus, 43
Consumer, rational expectations, 36,
37
Continuity equation, 60
Contract duration, 167. *See also* Price
discount
Corrosion. *See* Fatigue
Cost
amortization, 74, 83
elasticity, 71, 73
estimates, 56, 85
function, 82
incremental. *See* Incremental cost
minimization, 89
models, 87, 88. *See also* NAFCOM;
SSCM
of durability, 57. *See also* Marginal
cost

Cost (*cont.*)
   per day, 53, 74, 82, 85, 88, 90. *See also* Cost per unit time
   per payload, 82. *See also* Cost per transponder
   per transponder, 83, 88, 94. *See also* Cost per payload
   per unit time, 74. *See also* Cost per day
   profile, 70
   sink, 88
Cost, marginal. *See* Marginal cost of durability, Creative destruction, 114. *See also* Innovation; Obsolescence; Schumpeter, Joseph
Customers and durability choice, 45

Damage accumulation, 133
Deacon's masterpiece, 128. *See also* One-hoss shay
Decay, 6
Decay pattern. *See* Efficiency patterns
Decision-making, 29, 57, 82. *See also* Metrics
Degradation
  functional, 5
  physical, 5
Delta V., 69. *See also* Propellant budget
Demand imbalance, 97. *See also* Overcapacity
Depreciation, 34, 41, 110, 161
  accounting, 172
  and deterioration, 174
  and efficiency decline, 175
  and incremental present value, 178
  economic, 171
  empirical studies, 171

  patterns of, 185. *See also* Efficiency patterns
Depth-of-discharge, 65, 67. *See also* Battery, spacecraft
Design
  lifetime, 2, 5, 16, 101. *See also* Durability
  optimization. *See also* Metrics; Objective function to cost, 59, 82, 87, 101
Deterrence, strategy of, 42
Deterrent. *See* Barrier to entry
Discount rate, 104, 106, 108
Disposal, satellite, 69
DMSMS, 112. *See also* Obsolescence
Durability, 2, 5, 15
  academic interest in, 28
  and monopoly, 32
  and monopsony, 45
  and obsolescence, 42
  and quality, 22. *See also* Quality gap
  and service, 27
  and spacecraft, 62
  and value. *See* Value of durability
  capital intensity, 109. *See also* Durability, marginal cost of
  choice of, 7, 8. *See also* Durability specification
  competing on, 22
  competitive solution, 33
  demand for, 22
  ex ante, 16, 32, 46
  ex post, 32, 47
  extended, 16, 20
  history of (economic thought on), 24
  incremental. *See* Incremental durability. *See also* Incremental value

infinite, 7, 57
light bulb, 15
marginal cost of. *See* Marginal cost
    (of durability)
optimal, 59, 73, 103, 104, 105, 106,
    108. *See also* Durability, choice of
probabilistic approach, 48
reduced, 16, 17
specification, 7. *See also* Durability,
    choice of
spectrum, 8
staged choices, 48. *See also* Real
    options
technicalities of, 6
technically achievable, 7
underinvestment in, 121
Durable good, 8, 27

Eclipse operation, satellite,
    65
Economic life. *See* Service life
Economies of scale, 74, 83, 86
Efficiency patterns, 161, 185, 186. *See
    also* Depreciation, patterns of;
    One-hoss-shay goods
Environmental effects, 20. *See also*
    Waste
Equipment replacement, 110

Failure, 47, 57. *See also* Time to failure
Fatigue, 133
Flexibility, 136, 138
Flight Management System,
    140
Flow of service, 102
Ford, Henry, 130, 134. *See also*
    Model T
Functional degradation, 111
Functional inadequacy, 47

Game theory, 97
Gemini program, 5
General Electric, 30
Geostationary orbit, 150
Gillette, 31. *See also* Razor blade
    industry; Wilkinson
Gnomon, 1

Harvard, 135
Holmes, Oliver Wendell, 128, 129, 135.
    See also Deacon's masterpiece;
    One-hoss shay
Hotelling, Harold, 118, 173
Hubble Space Telescope, 89

Immortality, 4
Income statement, 173
Inconsistency, optimal plans, 50. *See
    also* Coase, Ronald; Consistent
    planning
Incremental
    cost, 73
    durability, 56, 73
    present value, 73
    value, 57
Initial operational capability,
    83
Injection errors, 68
Innovation, 20, 114. *See also*
    Obsolescence; Schumpeter,
    Joseph; Technological progress;
    Technology development
Integral design, 136. *See also* System
    architecture
Intel, 139
Internal combustion, 136. *See also*
    Benz, Karl
Inventory management, 110
Investment risk, 104, 108

Lamp life, 15, 30. *See also* Durability, light bulb

Landes, Charles, 4

Lastingness, 5

Lease price. *See* Transponders, lease price

Life degradation, 65

Lifespan, 5

Light bulb industry, 14, 31

Loading dynamics, 133, 146

Loading patterns. *See* Archetypes, loading

Lock-in, customer, 17

Longitude, spacecraft in GEO, 150

Losses, internalized, 38. *See also* Monopolist, renter versus seller

Maintainable systems, 62

Marginal cost (of durability), 36, 53, 56, 61, 74, 85. *See also* Competitive price

Market
  needs, 22
  power, 90
  share, 23, 91
  structure, 28
  valuation, 146

Mechanics, laws of, 3

Mercury program, 5

Metrics, 57, 81, 93, 97. *See also* Decision-making

Miller, Hillis J., 3

Model structure, 152

Model T, 130, 135. *See also* Ford, Henry

Modern ruins, 5

Modular design, 135. *See also* System architecture

Money, time value of, 74

Monopolist
  and pricing, 36
  and R&D, 42
  renter, 34
  seller, 34
  renter versus seller, 38

Monopoly power, 37, 40

Monopoly price, 37

MTTF [mean time to failure], 134

NAFCOM, 88. *See also* Cost models

New product introduction, 41, 42

Newton, Isaac, 3

Nietzsche, Friedrich, 4

Nondurable good, 8

NPV [Net present value], 103. *See also* Value

Objective function, 82. *See also* Design optimization; Metrics

Obsolescence, 15, 90, 92, 114, 137. *See also* Planned obsolescence; System obsolescence
  component, 114. *See also* Obsolescence, part
  in economics, 114
  in engineering management, 114
  onset of, 161
  overview, 111
  part, 112
  risk of, 20, 22, 74, 110, 116, 118
  time to. *See* Time to obsolescence

Oligopolists and durability, 34, 42. *See also* Cartel

One-hoss shay, 128, 129. *See also*
    Deacon's masterpiece
One-hoss-shay goods, 33, 185. *See
    also* Efficiency patterns
Opportunity loss, 20
Optimization. *See* Design
    optimization
Orbital parameters, 68, 73
Overcapacity, 90, 95, 96, 97
Ovid, 1

Parametric cost models. *See* Cost
    models
Partial derivatives, 108
Payload size, 97
Periodization, 26
Permanence, 5
Phoebus, 30
Planned obsolescence, 20, 42. *See also*
    Obsolescence; System
    obsolescence
Precommitments, 37, 41, 50. *See also*
    Coase, Ronald; Time
    inconsistency
Price
    discount, 167. *See also* Contract
        duration
    discrimination, 36
    elasticity, 73, 167, 168
    theory, 29
    war, 169
Price–quantity analysis, 28, 29,
    30
Pricing dynamics, 36
Product differentiation, 27
Product improvement, 17
Productive capacity, 185. *See also*
    Efficiency patterns
Propellant, 68, 69

Quality gap, 21

Razor blade industry, 31. *See also*
    Gillette; Wilkinson
Real options, 48, 123. *See also*
    Durability, staged choices
Reliability, 21, 132
Remote sensing, 118
Repeat purchases, 42
Replacement sales, 41
Replacement satellites, 157
Residual demand, 36
Retirement pattern, 172
Revenue generation, 105, 152
Revenue models, 89, 145, 150, 166. *See
    also* Utility profile
Risk
    and durability choice, 118
    of obsolescence. *See* Obsolescence,
        risk of
Risk-adjusted. *See* Discount rate
Risk-averse decision-maker, 119
Risk-neutral decision-maker,
    119
Risk-taking decision-maker, 119
Robust design, 142. *See also* Taguchi

Saint Augustine, 3
Sales stimulation, 17
Sales volume, 20. *See also* Sales
    stimulation
Satellite
    communications, 83, 85, 106
    cost per day, 83
    design lifetime, 22, 84, 106
    imagery, 118. *See also* Remote
        sensing
    loading dynamics, 153
    operators, 90

Schumpeter, Joseph, 114. *See also*
    Creative destruction; Innovation
Scaling effects, 61
Secondary markets, 21, 41, 43, 44
Sensitivity analysis, 107
Service contracts, 17, 21
Service flow, 89
Service life, 5, 47, 111
Service mix, 151
Shadow, 1
Shakespeare, William, 2, 4
Solar arrays, 65
Spacecraft subsystems, 62
SSCM, 88. *See also* Cost models
Stakeholders, 9
Station-keeping, 69. *See also*
    Propellant budget
Stochastic process, 153
Strotz, Robert, 49. *See also* Coase,
    Ronald
Substitution
    effect, 44
    perfect, 34, 44
Sundial, 1, 2
Sunk cost, 121–122
Supply-constrained market, 90, 93.
    See also Competitive market
Swan, Peter, 33
System
    architecture, 136. *See also* Integral
        design; Modular design
    obsolescence, 114. *See also*
        Obsolescence
    optimization. *See* Design
        optimization
    utilization, 146. *See also* Loading
        dynamics

Taguchi, 142. *See also* Robust design

Technological progress, 20. *See also*
    Innovation; Technology
    development
Technological slowdown, 22
Technology development, 34. *See also*
    Innovation; Technological
    progress
Telecommunications satellites, 22
Time, 1, 2, 3, 4, 5
    experience of, 2, 3, 5
    flow of, 3, 4
    inconsistency, 35, 36, 37, 39. *See
        also* Coase, Ronald
    measurement, 2
    to failure, 131, 136, 138. *See also*
        Failure
    to obsolescence, 91, 113, 114, 138
    value of money, 108
Time, uncertainty, and flexibility, 138
Trading period, 41
Transponders, lease price, 90, 96, 97,
    151
Transaction barriers, 41

Uncertainty and durability choice,
    118, 121–122
Usefulness, economic, 5
Utility profile, 88. *See also* Revenue
    models

Value, 88. *See also* NPV
    delivery artifact, 57, 89, 102
    forfeited, 121. *See also* Value, loss of
    incremental. *See* Incremental value.
        See also Incremental durability
    loss of, 22, 90. *See also* Value
        forfeited
    maximization, 89
    of a system, 23, 59

of durability, 57
Value-centric design, 87
Van Allen belts, 65
Volatility, 110

Warranty, 21

Waste, 21. *See also* Environmental
    effects
Wicksell, Knut, 27, 28
Wilkinson, 31. *See also* Gillette; Razor
    blade industry
Withrow, Gerald, 3